T0207462

Springer Biographies

The books published in the Springer Biographies tell of the life and work of scholars, innovators, and pioneers in all fields of learning and throughout the ages. Prominent scientists and philosophers will feature, but so too will lesser known personalities whose significant contributions deserve greater recognition and whose remarkable life stories will stir and motivate readers. Authored by historians and other academic writers, the volumes describe and analyse the main achievements of their subjects in manner accessible to nonspecialists, interweaving these with salient aspects of the protagonists' personal lives. Autobiographies and memoirs also fall into the scope of the series.

The books published in the Springer Biographies tell of the life and work of scholars, innovators, and pioneers in all fields of learning and throughout the ages. Prominent scientists and philosophers will feature, but so too will lesser known personalities whose significant contributions deserve greater recognition and whose remarkable life stories will stir and motivate readers. Authored by historians and other academic writers, the volumes describe and analyse the main achievements of their subjects in manner accessible to nonspecialists, interweaving these with salient aspects of the protagonists' personal lives. Autobiographies and memoirs also fall into the scope of the series.

Roberta Passione

Psychiatry and the Human Condition

A Scientific Biography of Silvano Arieti (1914–1981)

Roberta Passione
Department of Psychology
University of Milano-Bicocca
Milan, Italy

ISSN 2365-0613 ISSN 2365-0621 (electronic)
Springer Biographies
ISBN 978-3-031-09306-7 ISBN 978-3-031-09304-3 (eBook)
https://doi.org/10.1007/978-3-031-09304-3

Revised version of a translation from the Italian language edition: "La forza delle idee. Silvano Arieti: una biografia 1914-1981" by Roberta Passione, © MIM EDIZIONI SRL 2020. Published by Mimesis.

Cover illustration:
Front cover photo: Silvano Arieti, 1976. Previously published in Psychotherapy of Schizophrenia and Dynamic Cognition by Silvano Arieti. Reprinted with permission from Roche
Back cover photos: Silvano Arieti, circa 1937; Silvano Arieti, circa 1940. Courtesy of James Arieti—All rights reserved

This Springer imprint is published by the registered company Springer Nature Switzerland AG
The registered company address is: Gewerbestrasse 11, 6330 Cham, Switzerland

Preface

This book is the result of research in the archive of Silvano Arieti (Silvano Arieti Papers, hereafter SAP), held by the Manuscript Division of the Library of Congress, Washington, D.C., where I delved into published and unpublished sources. The papers are now arranged at the Library in five series, but during my research, cataloguing the collection was still in progress. The series entitled *Correspondence*, for example, had not yet been divided into subseries, and for this reason I refer to the general series *Correspondence* without mentioning the subseries in which a document may currently be found.

All references are listed at the end of the chapter in which they occur. Archival sources are listed as SAP and include year, type of source (letter, manuscript, typescript, handwritten notes, etc.), further details of the document when available (exact date, author, title and/or topic, etc.) and the reference to the series in which the document has been catalogued.

For example; SAP (1940). L, July 12 (Silvano Arieti to Robert M. Yerkes). In *Correspondence, 1940–1981* [stands for Silvano Arieti Papers (SAP), year (1940), type of source (L for letter), exact date (July 12), author, title and/or topic (Silvano Arieti to Robert M. Yerkes), series in which the document has been catalogued (*Correspondence, 1940–1981*)].

Undated archival documents are listed as SAP (U) [Undated]. For example: SAP (U). T (Silvano Arieti, *The Conception of Man*). In *Speeches and Writings, 1940–1981. Articles. Undated* [stands for Silvano Arieti Papers (SAP), year (Undated), type of source (T for typescript), author, title and/or topic (author Silvano Arieti, title *The Conception of Man*), series in which the document has been catalogued (*Speeches and Writings, 1940–1981. Articles. Undated*)].

In cases where a series contains several references concerning the same issue, I provide neither a year nor cite a particular document but I refer only to SAP and the name of the series.

For example; SAP. *Correspondence, 1940–1981.*

Also, I used only SAP without mentioning a particular series for documents that had not yet been catalogued in any series during my work at the Library.

Archival sources from the Historical Archive of the University of Pisa, Italy, are listed as HAUP.

A list of abbreviations may be found on page xi

Because of the biographical nature of this book, I chose a narrative style, as I believe that a life should not be forced into a merely academic mould and that a rigidly donnish perspective would be inimical to the very purpose of this biography—bringing science back to the life from which it stems. Nevertheless, I wish to emphasize that every detail of this volume is based on documentary sources and that the book contains no fictional material.

Milan, Italy Roberta Passione

Acknowledgements

In writing this book, I have been helped in different ways by many people. My heartfelt thanks go first to James Arieti for helping me with the translation of my manuscript into English. His generous help has been invaluable for my work. I also wish to thank the Manuscript Division of the Library of Congress and, in particular, Jeffrey Flannery and Michael Folkerts, who have kindly welcomed and assisted me during my research on Silvano Arieti's papers. I am grateful also to Marco Bacciagaluppi, with whom I had many conversations, and who honoured me with his attention since I began my work on Arieti.

Many others have helped, and though I shall simply list them, my gratitude to them is genuine and hearty: Mauro Antonelli, David Arieti, Valeria Babini, Rita Bruschi and the Silvano Arieti Association of Pisa, Massimo Bucciantini, Andrea Castiello d'Antonio, Alessandra Cerea, Guido Cimino, Marco Conci, Eugene Donovan, Angelo Fioritti, Pier Francesco Galli, Patrizia Guarnieri, Pompeo Martelli, Andreas Mayer, Aurelio Molaro, Paolo Peloso, Bruna Zani and the Gian Franco Minguzzi Institution of Bologna.

I am grateful to those friends and relatives who sustained me over the past difficult years. I cannot list them all, but I wish to single Alberto, whose love is always restorative; and Michele, Bruno, and Demetrio, who always bring me joy and make me smile.

This is the first book that I am writing without my mother living in this world. She was always my first reader, and I shall always miss her comments and advice. With deep-rooted love, I dedicate this book to her memory.

Contents

Abbreviations

B	Brochure
C	Certificate
F	Form
HAUP	Historical Archive, University of Pisa, Italy
HN	Handwritten Notes
IC	Invitation Card
L	Letter
M	Manuscript
P	Poem
PB	Promotional Brochure
PD	Personal Document
Ph	Photograph
SAP	Silvano Arieti Papers, Manuscript Division, Library of Congress, Washington D.C.
T	Typescript
TL	Telegram
TMin	Typewritten Minutes
U	Undated
UA	Unsigned/Unidentified Author
Unt.	Untitled
UR	Unidentified Recipient
US	Unsigned/Unidentified Sender

Chapter 1
Introduction

> The point is that it is stories about people, not the search for analogies to the periodic table of elements and fundamental particles in physics, which constitute what we really want to know about human beings (Young 1988, p. 113).

Biography, described as a basic discipline for human science, is an essential genre in the history of science. It enables us to overcome positivistic and naive conceptions of science by shedding light on the human origins of scientific knowledge. As a "key to epistemology in action" (Young 1988, p. 122), it focuses on the role played by subjectivity and thus it counteracts scientism—the "ism" described by Félix Le Dantec as the science that retains no trace of its human origin (Gusdorf 1960).

At the same time, biography provides insights not only on individuals but also on collective processes; with the immediacy of specific cases, it brings to the surface core questions in history—the relationships between freedom and necessity, general and particular, and singular and plural (La Vergata 1995; Shortland & Yeo 2008). In this way a biography can help us understand both a scientist and his discipline. Keeping this principle in mind, in the following pages I explore the life and work of Silvano Arieti, trying to draw some insights about the development of psychiatry in the twentieth century.

Throughout his professional life, Silvano Arieti carried out a study of schizophrenia, work that received wide recognition. Yet, with only a few exceptions (Bemporad 1981; Bacciagaluppi and Bacciagaluppi Mazza 1982; Alger 1983; Ciani 1984; Bruschi 2001; Bruschi 2002; Silver 2005), an historical account and a deep analysis of his full scientific contribution has been neglected. He was more than a famous psychiatrist awarded the National Book Award for the second edition of *Interpretation of Schizophrenia*, published in 1974 (Arieti 1974). He was a profound thinker who promoted an original epistemological approach to the scientific study of human beings. For Arieti, psychiatry was a thorough investigation of the human condition, a condition characterized by a basic dialectic of similarity and difference. In this way, Arieti conceived the exploration of mental illness as a means to understand that "the other" actually resembles ourselves and is a constituent part of our identity (Passione 2016a, 2016b, 2019).

R. Passione, *Psychiatry and the Human Condition*, Springer Biographies,
https://doi.org/10.1007/978-3-031-09304-3_1

Influenced by the teachings of Harry Stack Sullivan and of the interpersonal school, Silvano Arieti made this issue of similarity and difference the focus of his psychiatric and scientific work. His intellectual arc calls to mind what Tzvetan Todorov wrote about "the discovery of *the other*" in the conquest of America and the role played by this discovery in the development of our modern Western identity:

> We can discover the other in ourselves, realize we are not a homogeneous substance, radically alien to whatever is not us: as Rimbaud said, *Je est un autre*. But *others* are also "*I*"s: subjects just as I am, whom only my point of view ... separates and authentically distinguishes from myself ... The history of the globe is of course made up of ... discoveries of others; but ... it is in fact the conquest of America that heralds and establishes our present identity; even if every date that permits us to separate any two periods is arbitrary, none is more suitable, in order to mark the beginning of the modern era, than the year 1492, the year Columbus crosses the Atlantic Ocean (Todorov 1983, pp. 3–5).

Todorov's reference to America is highly suggestive in the case of Arieti, for his encounter with the New World represented a turning point in his life—the beginning of his personal and scientific journey (Passione 2021).

Following the promulgation of Italian racial laws, in January 1939 Silvano Arieti left Pisa, his hometown in Italy, to escape the anti-Semitic legislation. He crossed the ocean to settle in the United States, and there his encounter with "the other" began—first through confronting the challenges of a foreign country and an unknown language, then through his experience with schizophrenic patients at Pilgrim State Hospital, the largest psychiatric hospital in the world.

From early in the life of Arieti one can perceive a close connection of science and his experiences. The choice of psychiatry, for example, was not the result of an adhesion to the positivistic "religion of science" that had led many colleagues from a previous generation to study medicine (Passione 2007) but stemmed from an actual involvement—his relationship with Pardo Roques, president of the Jewish community of Pisa, who suffered from a severe phobic disorder (Arieti 1979).

The fusing of personal events and scientific work emerges also in the main fields of Arieti's clinical practice, starting with the phobic disorder afflicting his very first patient (Arieti 1979). Subsequently, schizophrenia excited his interest when he arrived at Pilgrim State Hospital for a job as an inexperienced intern not yet proficient in his new language—almost an *alien* among aliens in "the largest psychiatric hospital in the world" (Arieti 1978, p. ix). Later, he devoted his attention to depression (Arieti & Bemporad 1978), which he had experienced within his family (SAP).

I wish to emphasize, however, that with this volume I am not advocating psychohistory, a subject to be studiously avoided, as Leon Edel, the biographer of Henry James, suggested in 1961 (Edel 1961). Instead, I invite the reader on a twofold exploration—of the life of an individual, through his published and private papers, and of the history of twentieth-century psychiatry, with its debates, contradictions, and unsolved problems.

Spanning a period from 1914 to 1981, Arieti's life ranges over two continents and encompasses nearly half a century of the history of psychiatry. Educated in Pisa, Italy, at the neuropsychiatric school of Giuseppe Ayala, and in the United States at the William Alanson White Institute, in New York, Arieti combined a meticulous scientific study of human behavior with a wide-ranging meditation on the human condition. He aimed at integrating knowledge from different disciplines and from contrasting perspectives. In doing so, Arieti never tried to avoid perplexities or contradictions but invited his colleagues to acknowledge them as an inherent part of their work, or, more generally, as an inherent part of the scientific study of human beings.

"There is no easy road to love," he wrote in 1977 with his son James (Arieti & Arieti 1977, p. 5); we can say the same for his conception of psychiatry and human science at large—there is no easy road to knowledge of human beings. In his last years he looked with concern at the increasing success of various kinds of reductionism (in both socio-cultural and biochemical-genetic approaches). He considered one-sided views and partisan warfare between opposing paradigms to be an impoverishment of psychiatry. At the end of the 1970s, the success of the *Diagnostic and Statistical Manual of Mental Disorders* (DSM) marked a turning point for psychiatry. With the standardizing and non-theoretical approach of the *DSM*, Arieti's voice would end up simply another elegant but useless and outdated example of scientific reasoning in psychiatry.

Perhaps it is time to listen to that voice again.

References

Alger, I. (1983). The Intellect and Humanism of Silvano Arieti. *Journal of the American Academy of Psychoanalysis* 11 (1), 15–34.

Arieti, S. (1974). *Interpretation of Schizophrenia.* Basic Books.

Arieti, S. (1978). *On Schizophrenia, Phobias, Depression, Psychotherapy and the Farther Shores of* Psychiatry. Brunner Mazel.

Arieti, S. (1979). *The Parnas.* Basic Books.

Arieti, S., Arieti, J. (1977). *Love Can Be Found.* Harcourt Brace Jovanovich.

Arieti, S., Bemporad, J. (1978). *Severe and Mild Depression: The Psychotherapeutic Approach.* Basic Books.

Bacciagaluppi, M., Bacciagaluppi Mazza, M. (1982). An Italian Commemoration of Silvano Arieti. *Academy Forum* 26 (1), 9–10.

Bemporad, J. (1981). In Memoriam. Silvano Arieti 1914–1981. *Journal of the American Academy of Psychoanalysis* 9 (4), III–VII.

Bruschi, R. (Ed.), (2001). *Uno psichiatra tra due culture. Silvano Arieti (1914–1981).* ETS.

Bruschi, R. (Ed.). (2002). *Continuare senza dimenticare. Silvano Arieti (1914–1981).* ETS.

Ciani, M. G. (1984). Ritratti critici di contemporanei. Silvano Arieti. *Belfagor,* 39 (6), 645–663.

Edel, L. (1961). The Biographer and Psycho-Analysis. *International Journal of Psychoanalysis,* 42, 458–466.

Gusdorf, G. (1960). *Introduction aux sciences humaines.* Le Belles Lettres.

La Vergata, A. (Ed). (1995). *Le biografie scientifiche.* Il Mulino.

Passione, R. (2007). *Ugo Cerletti. Il romanzo dell'elettroshock.* Aliberti.

Passione, R. (2016a). Fra storia e natura: la psichiatria antiriduzionistica di Silvano Arieti. *Ricerche di psicologia,* 39 (2), 152–179.

Passione, R. (2016b). La psichiatria di Silvano Arieti: un primo profilo. *Physis. Rivista Internazionale di Storia della Scienza,* 51 (1–2), 219–330.

Passione, R. (2019). "A Glimpse of Inner Struggles": Psychiatry and Human Identity on the Work of Silvano Arieti. *European Yearbook of the History of Psychology,* 5, 83–109.

Passione, R. (2021). Silvano Arieti (1914–1981). In P. Guarnieri (Ed.), *Intellectuals Displaced from fascist Italy. Migrants, Exiles and Refugees Fleeing for Political and Racial Reasons.* Firenze University Press. http://intellettualinfuga.fupress.com/en

SAP. *Correspondence, 1940–1981.*

Shortland, M., Yeo, R. (Eds.). (2008). *Telling Lives in Science: Essays on Scientific Biography.* Cambridge University Press.

Silver, A.L. (2005). Silvano Arieti. Remembering His Message. *PsiLogos,* 2 (1), 23–38.

Todorov, T. (1983). *The Conquest of America. The Question of the Other.* Harper & Row.

Young, R. (1988). Biography. The Basic Discipline for Human Science. *Free Association,* 11, 108–130.

Chapter 2
Castaway

2.1 The Storm

April 1976, Manhattan, New York City. The phone rings in a medical office at 125 East 84th street. *You've reached the doctor's voicemail: please, leave a message.* At the other end of the line, Paul Miller, a young student at the Insight Counseling Center of Niagara Falls, N.Y., is taken aback and hangs up. Then, changing his mind, he dials the number again and spells out to the receiver the reason of his call. He points out that it is not an emergency call and that he is not looking for a visit; he would just like to talk with the doctor in order to receive some biographical information about him (SAP 1976c).

The doctor is Silvano Arieti, a psychiatrist of Italian origin. Just one year earlier he had won the National Book Award in Science for *Interpretation of Schizophrenia* (Arieti 1974), a groundbreaking work of psychiatric literature. Reading this book elicited in the young Miller a great deal of thinking about his own future role as a therapist, as well as a desire to learn more about this famous psychiatrist. So, he decided to prepare a short paper on Arieti, but quickly discovered that searching libraries and bookshops for biographical material was a hopeless undertaking: he could find nothing about Arieti, except for the "brief blurbs on the dustjackets of his books" (SAP 1976c).

The following afternoon, Paul Miller receives a call from Silvano Arieti's secretary, Joan Kirtland. She says that the doctor is available to talk at 4 p.m. Miller waits impatiently; then, at the indicated hour he dials the number. Arieti himself answers the phone, a voice with a gentle continental accent. He is very kind and sounds pleased and surprised at Miller's idea: "Maybe it's not good for my ego, but to my knowledge nobody has ever written a biography" (SAP 1976c). Arieti then draws Miller's attention to *The Will To Be Human,* a book he published in 1972 (Arieti 1972), which contains some biographical information. It might be useful, he suggests, to start at the point in the book where he describes a conflict with his father, which he considers a turning point in his life: "I had been rather brainwashed by the Italian fascists as a young boy, and my father was horrified. I think this argument

R. Passione, *Psychiatry and the Human Condition*, Springer Biographies,
https://doi.org/10.1007/978-3-031-09304-3_2

was very important in my development. I am a Sephardic Jew and my family settled in Italy following the Spanish Inquisition, and have lived there … until I emigrated to the United States at twenty-four" (SAP 1976c).

The following May, Arieti finds an envelope on his writing desk: it is Miller's typescript, accompanied by a letter in which Miller announces his intention to turn his paper into an article and to look for a journal willing to publish it (SAP 1976a).

The short paper, evocatively entitled *An Interpretation of Arieti*, is a biographical note on the theme of conflict as a pathway to personal identity. In Miller's thinking, the tale of a democratic Jewish doctor's son, torn between paternal authority and the fascist slogan "Believe! Obey! Fight!," goes beyond Arieti's individual life to find the story's roots in the Biblical accounts of the prodigal son, Cain and Abel, and Jonah. Included in Miller's pages is a reference to another famous confrontation between father and son in contemporary literature: the one that takes place in *Portnoy's Complaint,* by Philip Roth, published in 1967 (SAP 1976c; Roth 1967).

In Miller's interpretation, the conflict could have been resolved only after Silvano Arieti's emigration to the United States—an unknown new world when the disoriented and bewildered boy could finally experience a different sense of his roots and look with new eyes to paternal authority. In conclusion, figuratively speaking, it was necessary for him to experience a shipwreck in order to return home.

"Dear Mr. Miller," Arieti replies on May 25, "I was very much impressed and honored by your paper … I must say that some parts have helped me to understand myself better. You say that you want to do a biographical piece on me. Of course … I'll give you all the assistance you need, but I doubt very much that you'll find somebody who wants to publish such an article" (SAP 1976b).

This pessimism had its reasons, as another attempt had already ended in nothing a few years before. In 1969, the journal *Voices* had planned to publish a biographical sketch on Silvano Arieti in a special issue on "Contemporary Masters of Psychotherapy." The manuscript, however, met the expectations neither of Arieti nor of the editor, Vin Rosenthal, for it was a cold and formal tribute to the theorist and therapist, but revealed nothing about the *person* and the life of the subject (SAP 1969a, 1969b, 1969c, 1970).

In those very same days, a more personal biography had been written by Silvano Arieti's wife, Marianne Thompson. Because of her bond with her husband, it would not have been an appropriate decision to publish it as a contribution on contemporary masters of the field. It appeared, however—quite under the radar—in the *Newsletter of the Society of Medical Psychoanalysts* (Thompson Arieti 1969).

Setting out to write an intellectual and scientific biography of this psychiatrist, I shall follow the line traced by Marianne Arieti, who, in her concise yet vivid contribution, casts her eyes on Silvano as a child. I too shall begin from his boyhood and youth in Pisa, trying to depict him both on the basis of published writings and, more fully, on the basis of the archival documents held by the Manuscript Division of the Library of Congress in Washington, D.C.

Silvano Arieti was born in Pisa on June 28, 1914—that is, the same day as the assassination of the Archduke Ferdinand which unleashed World War I, a coincidence that Arieti would always indicate as the sign of a life "forever linked with the larger human drama" (Alger 1983, p. 16). His father, Elio Arieti, was a general practitioner. Elio was born out of wedlock on February 26, 1888. Under the name Elio Rosi, he was wet-nursed for three years. Then, in 1891, after his parents' marriage, he was acknowledged as their son by Argia Fumaioli and Vittorio Arieti—who preceded his son as a physician (SAP 1888).

On December 15, 1912, Elio Arieti married Ines Bemporad (SAPf), who had been born into a merchant family (Fig. 2.1).

Ines' father, Giacomo Bemporad, was a highly valued figure in Pisa's Jewish community. He had arrived as a peddler from Piedmont, and in Pisa succeeded in building a chain of prosperous stores (Arieti 1979a). Silvano Arieti loved spending his afternoons as a child in one of these, located in the center of the city, and run by his grandmother. For him "this store was a living theater filled with characters in search of a would-be psychiatrist: the usurer, the alcoholic, the mother of prostitutes, the peddler, the pauper, the derelict, as well as the affluent and the aristocratic" (Thompson Arieti 1969, p. 43).

Elio Arieti raised his two sons, Silvano and a younger brother Giulio (Fig. 2.2), with severity and austere benevolence. He had a very critical attitude toward the era in which he lived, considering it an age that was "bringing the man back to the caves"

Fig. 2.1 Ines Bemporad and Elio Arieti, circa 1925 (Personal Collection of James Arieti) © Courtesy of James Arieti—All rights reserved

Fig. 2.2 Silvano and Giulio with Ines Bemporad, *circa* 1935 (Personal Collection of James Arieti)

(SAP Ub, my translation) rather than leading it forward. Therefore he tried to convey other values to his children: justice, freedom, and brotherhood, which, by means of culture, are capable of raising "Human Thinking" (SAP Ub, my translation) above its current coarseness.

Silvano was also initiated into this "human thinking" at the elementary school of the Jewish congregation, where, along with reading and writing, he learned a simple, pristine, and clear way of thinking that he would later credit to his beloved teacher, Luisa Orvieto (Arieti 1979a, 1976).

Arieti's school years coincided with turbulent political times. He went to live with his maternal grandparents, where "around the large dinner table he heard lively debates" among his uncles and aunts, numerous friends and relatives, "on how to deal with Fascism" (Thompson Arieti 1969, p. 43). It was an animated and warm family environment, where the young boy could find motivation, attention and support, especially from his Aunt Yolanda and her husband, Giampaolo Rocca, who used to walk him to school on his first days of classes (Arieti 1979a).

On the other hand, his relationship with his father was not so smooth as with his mother's family. Remaining aloof from the political scene, which he detested, Elio Arieti had an attitude of adamant silent protest, which he conveyed in his daily life and work with a true democratic spirit, always to the benefit of the poor and the needy. More importantly, he felt a desperate aversion to Fascism, whereas the young Silvano did not. The pretentious rhetoric of Fascism, with its idealizing taste, exerted a powerful grip on him: "Mussolini became the hero…for me too, when I was ten to thirteen years old" (Arieti 1972, p. 86). At the age of thirteen Silvano wrote a poem in honor of the Duce, and had the rebelliousness, or perhaps just the recklessness, to declaim it at the cultural club of the Jewish community. Elio grieved, and, unable to accept that the fascists had succeeded in brainwashing his son, reprimanded him harshly.

Here we come to the conflict that Arieti indicated to Miller as a pivotal issue of his existence—the sort of conflict with the authority and his own origins that the young Silvano conveyed also in his numerous tales. For example, in a story entitled *Il Rabbino di Kleinay,* he writes of Sara, a young Jewish girl, who collides with her father because of her love for a Christian boy (SAP Ui). Like Sara, who betrays her people by falling in love with a Catholic, Arieti betrays his father by "falling in love" with Mussolini.

In 1929 Silvano Arieti enrolled at the "Liceo Galilei," where he obtained his high school diploma in classical studies on September 26, 1932. Thin and frail (he was exempted from gymnastics), he loved studying (SAPc; SAPd). He was enthralled by the Greek classics and the tragedies of Shakespeare. Pirandello fascinated him; Dante enchanted him—especially in the imaginative and dreamlike world of the *Vita Nova.* Vico's *Scienza nuova* captivated him. These were his early steps toward the study of the human mind: "long before psychiatry came of age, philosophers, poets, artists and playwrights had deep insights into the human phenomenon of madness" (SAP 1973a).

Young Silvano himself (Fig. 2.3) was also very keen on writing. He filled notebooks with tales and screenplays that reveal a dreamy and a spiritual nature, "thoroughly full of love" (SAPd, my translation) and torment, inclined to melancholy and a certain tragic vision of existence as well as to bursts of enthusiasm and joy.

Shy and prone to stomach aches (SAP 1961),[1] he suffered from "a certain hesitation" that was "part of [his] personality" (SAP Ug, my translation); he had a solitary and an introspective attitude—almost like a hermit (SAP Uj; SAP Uk); yet he was sociable, often surrounded by friends. He was a combination of opposites found also in the characters of one of the stories he wrote when he was sixteen years old: the melancholic, taciturn, and introverted Saverio; and Luciano, an optimist with a sunny personality, always ready to make new friends (SAP 1930). These characters embody those polarities of the human spirit that eventually will attract his attention as a psychiatrist; but long before he became a psychiatrist, he showed traits of a young man "most complex and full of contradictions, … of extraordinary complexity and extraordinary simplicity" (Clemmens, Spiegel, Bieber & Di Cori 1982, p. 8).

In a sense, it is easy to notice a certain continuity between the interpersonal assumptions that will later characterize Arieti's psychiatric work and the youthful voluble Arieti, who spent much of his time in animated discussions with his friends.

[1] In 1961 his father wrote from Italy: "We are so glad that you're ok with your stomachache. I remember that also when you were at school I used to give you some antacid pills".

Fig. 2.3 Silvano Arieti as a young man, *circa* 1932 (Personal Collection of James Arieti) © Courtesy of James Arieti—All rights reserved

He pursued dialogue even in the pages of his notebooks. For example, in a poem he dedicated to his friend Vittorio Righi, who was enthusiastic about the advances in technological progress—an enthusiasm that Silvano did not share:

> It's not because I disregard human knowledge, but to comfort my soul I look elsewhere. When it's summer, and thousands of stars shine in the sky ..., when it's spring and you see new wonders blooming from every branch, lighting up your eyes, your senses, and your whole heart; when in the country and amidst the woods you can smell the green grass and feel a freer and more beautiful sense of life ... it's there, Vittorio, that I find comfort and peace. And then, satisfied, with my head bent, I think deeply, and my thankfulness rises high, up to God (SAP Ul, my translation).

With these words Arieti describes something that can help us to understand his move toward medical studies: his choice did not stem from a positivistic religion of science, nor from an enthusiasm for futuristic progress, but from a religion of nature that in its tragic character has almost a romantic taste. It is a nature that produces wonders but also harsh laws. In a poem dedicated to his father, Arieti dwells upon the common destiny of death that characterizes everything living in this world, where nature is a "superhuman intelligence" (SAP Uc, my translation) constantly threatening our life, relentlessly drawing us to death. Nobody can dodge its laws. This conception of the world helps us to understand Arieti's choice of pursuing medicine as a career, a choice that reminds us of the conception of the French "philosophe" Pierre Jean Georges Cabanis about the source of medical science—the instinct to help those who suffer (Cabanis 1798; Moravia 1974; Staum 2016). Addressing his father, Arieti writes: "Guilty is nature, but holy is the regiment to which you belong, which in the name of science brings a ray of hope to man and comforts him with the illusion of pushing away a little the mystery of death and of the unknown" (SAP Uc, my translation).

In 1932 Arieti enrolled at the medical school of the University of Pisa (HAUP; Arieti 2001). He moved from his philosophical and literary interests to anatomy, which "absorbed, paralyzed, and drowned all [his] other ideas" (SAP Uf, my translation). He was an excellent student. He was fond of clinical medicine and physiology, but most of all of psychiatry. In fact, for a passionate writer of inner turmoil, psychiatry was the branch of medicine that most resembles writing a drama (Arieti 1999). In addition, we must pay attention to the impact exerted on him by Pardo Roques, president (Parnas) of the Jewish community, whose salon in Pisa Arieti began to frequent the same year that he enrolled at the University. Roques—whose story Arieti will tell in 1979—suffered from a severe phobic disorder that did not allow him to go far from home. It was Roques' inner turmoil—the contrast between his high intellectual, moral stature and the miserable slavery to which his illness condemned him—that most touched Arieti, contributing to his choice of medical specialty: "I intended to become a psychiatrist because of him … His illness had an aura of mystery that I hoped I could unveil one day" (Arieti 1979b, pp. 8, 16; Fig. 2.4).

Thus, Arieti attended the Clinic of Nervous and Mental Diseases headed by Giuseppe Ayala (HAUP; SAPc), a pupil of the Italian neurologist Giovanni Mingazzini (Triarhou 2021). Ayala introduced Arieti to the work of Kurt Goldstein and an avantgarde neurology (Arieti 1959).

Fig. 2.4 Silvano Arieti during his university years, *circa* 1937 (Personal Collection of James Arieti)

Still, according to the traditional biological focus of Italian psychiatry (Babini 2009, 2014), Arieti's university education was deficient in psychological and dynamic training. In 1973 Arieti wrote:

> Still vivid in my mind is a demonstration given by my teacher of psychiatry and neurology while I was a student in medical school. Incidentally, my teacher was an excellent neurologist whose name is prominently mentioned even in present day textbooks. ... In the lesson I am referring to he was presenting to the class a catatonic patient who was in a statuesque position except for an "incongruous" smile which occasionally appeared and a movement of an arm repeated rhythmically from time to time. After demonstrating the classic catatonic features ..., my teacher said to the patient: "Giuseppe, I have just received a telegram announcing that your mother has suddenly passed away." The patient did not wince. He continued to have his incongruous smile ... and to move his arm ...
>
> "Do you see," said the teacher, "how insensitive he is? He is not even affected by the news of the death of his mother." The whole class, myself included, was spellbound. What was so astounding at that time was not so much the callousness with which the news of the alleged death of one's own mother was announced, but the fact that a son would remain impassive at such news. How was that possible? Was the patient still a human being? ... Of course, we know better now. In spite of the catatonic withdrawal, the patient ... was not unfeeling; he was sensitive and anxious to the point that he had to cut almost all the ties with a world which, in practically all its communications, caused him so much fear, pain, was so callous and cruel. Now, almost four decades later, the greatest cruelty in this episode seems to me to reside in the fact that the professor did not know he was cruel. On the contrary, he wanted to help, he was motivated by his usual didactic fervor. As I have already mentioned, he was an excellent neurologist, a good man personally, but he lived and taught at a time when professors of psychiatry knew very little about psychodynamics (Arieti 1973a, pp. 338–339).

Therefore, Arieti began an independent venture into the field of psychoanalysis, which he approached through the writings of Edoardo Weiss and the work of Enzo Bonaventura, whom he met in person at Pardo's cultural gatherings, where for the first time he heard about Freud (Arieti 1979a).

Ayala, to whom Arieti revealed an interest in psychoanalysis, recognizing his student's enthusiasm and skill, allowed him to approach some patients at the university psychiatric clinic, in order to put his readings to the test. In this way Arieti met Pietro, his first patient upon whom he tried a psychological therapy (Arieti 1979a).

On July 15, 1938, at 4 p.m., Silvano discussed his thesis entitled *Contributo allo studio delle encefalomieliti* ("A contribution to the study of encephalomyelitis") and was graduated with honors in medicine (HAUP). Nevertheless, the joy at this brilliant achievement was irreparably marred the day before, with the publication in the *Giornale d'Italia* of an article entitled "Il Fascismo e i problemi della razza" ("Fascism and the Problems of Race"), the repugnant content of which paved the way for the racial laws of the following autumn.

"Jews do not belong to the Italian race." Who can imagine how many times Arieti read this menacing passage "printed in large letters" (Arieti 1979a, p. 96) in the article, marking a tragic turning point in his life? The full storm was about to arrive. The blow more terrible for him personally, since he used to trust Fascism. Until shortly before, he had also been a member of the University's fascist organization (HAUP). In one photo we can see him approach a stage during a muster, wearing a uniform and holding a paper in his hand, perhaps to read from it (SAPj).

Fig. 2.5 Silvano Arieti's passport, cover page, 1938 (SAP, Library of Congress, Manuscript Division) © Courtesy of James Arieti—All rights reserved

From this point on, everything changed. At home Uncle Giampaolo and Aunt Yolanda tried to convince their relatives to leave the country, but Elio and Ines were unwavering in their decision to remain in Italy. They nevertheless wanted their sons to save themselves and leave.

Like Angelo, one of the characters of his book on the Parnas, Silvano was deeply torn (Arieti 1979a; SAP 1979c). He did not want to leave Italy but, at the same time, he could not contain his growing fear of the escalating anti-Semitic legislation that had been set in motion. For this reason he spent a terrible summer, during which he continued to attend the Hospital "Santa Chiara" of Pisa, where he had started his internship shortly before graduation. There, between abscesses and contusions, fistulas and broken bones, hernias, gastric soundings, and spinal taps (SAP 1938a),[2] he made up his mind. In September, Ayala wrote a reference letter for him that certified his aptitude for neuropsychiatry (SAP 1938b). Arieti applied for a passport, which was granted on December 31, 1938, and bore the photo of a young and thin Silvano (Figs. 2.5 and 2.6). France and Switzerland were permitted destinations, but England and United States were not (SAP 1938c).

On the cold night of January 11, 1939, Silvano Arieti kissed his mother and father goodbye and escaped from "the hordes of Mussolini" (Arieti 1968, p. 86). Everything took place in great secrecy (he was officially going on a skiing holiday)

[2] A list of the medical procedures performed by Arieti is reported in the daily journal of Arieti's internship at the Santa Chiara Hospital in the academic year 1938–1939.

Fotografie

Spazio riservato alla fotografia della moglie

Firma del Titolare

Silvano Arieti

Autentificazione della firma

Silvano Arieti

IL PODESTÀ

Firma dell'Autorità

31 DIC. 1938 Anno XVII

della moglie

Fig. 2.6 Silvano Arieti's passport, inside page, 1938 (SAP, Library of Congress, Manuscript Division) © Courtesy of James Arieti—

and in a hurry: he did not even have time to take the State examination to practice medicine, for which he had applied (SAPc; SAPf).

In Switzerland, his passport bore the stamp *Ville de Lausanne,* where Arieti attended the university psychiatric Clinic (Hôpital de Cery) for one month (SAP 1939a), a sojourn he would remember "with gratitude and affectionate nostalgia" (SAP 1971). Switzerland became for him a momentary oasis, *The Swiss Oasis* about which he would one day write a play under the pseudonym of Abramo Tuscany (SAP 1973b). Perhaps, physically and mentally, Switzerland was a country not so distant from his home, a place where he could still find something familiar—the French language, for example, which he had studied in school, as distinct from English, which he did not know at all (SAP 1975). During his stay in Lausanne he made the acquaintance of four Polish students. Three of them would perish in the concentration camps (SAP 1965).

Lausanne, however, was just a brief interlude, as he waited for a temporary visa which would enable him to reach Great Britain, where he had to obtain a second visa for the United States. Silvano arrived in London on March 10, 1939. Ten days later he obtained permission for intercontinental expatriation. On Saturday, April 1, he boarded the Queen Mary for New York. The ship was commanded by Commodore Irving and Captain Snow. On the cover of the travel brochure was a pale blue globe silhouetted against a deep blue sky. The coastline of the two continents was outlined, London and the Big Ben on one side, the New York skyline on the other (SAP 1939b).

Silvano traveled with his brother Giulio, his Uncle Enrico Bemporad and Aunt Vana, and their two children—Giulio (Jules) and Giacomo (Jack), ages 1 and 5 respectively (Passione 2021). Yolanda and Giampaolo, who had already left, were waiting for them in America—a precious outpost in the storm.

2.2 Stay the Course

The Queen Mary docked in New York on April 6, 1939. With his relatives, Silvano disembarked on Ellis Island, the epic gateway to the United States for millions of people coming from all over the world. Immigrants were received in the great Registry Room. The inspection process lasted many hours, with a document check, medical examination and mental testing (Bateman-House & Fairchild 2008).

Today, Ellis Island houses the National Museum of Immigration, where a rich exhibition brings the visitors close to the lives of those who used to crowd this place. Mental tests are among the historic records that have been preserved. During a visit at the Museum a few years ago, the writer of this book noticed one particularly suggestive test—a simple card asking the examined person whether in washing a flight of stairs it was better to start from the highest or lowest step. An answer, given in perfect English by a twelve-year-old girl, is impossible to forget: "*I did not come to the United States in order to wash stairs.*" It could have been Silvano Arieti's answer, too, if only had he been able to read the future, and if he had known English.

No news can be gathered from archival documents about early days of Arieti in the United States. We can only imagine his confusion and disorientation, as he tried to decipher an unknown language in which he "never had a single lesson in (his) whole life" (SAP 1975). What we do know is that in the U.S. Giulio and Silvano came to the parting of their ways: the younger Giulio went to Miami, where their Aunt Yolanda had started a business, and the older brother stayed in New York, where Ayala had recommended him to Armando Ferraro, an Egyptian-born neuropathologist who trained in France, under Pierre Marie, and in Italy, under Luigi Luciani and Giovanni Mingazzini. Liberal and antifascist, Ferraro was Chief of the Department of Neuropathology at the New York State Psychiatric Institute (Roizin 1983). Most importantly, he knew Italian.

Ferraro welcomed Silvano Arieti with kindness and thoughtfulness, and helped him to receive a grant in neuropathology from the Dazian Foundation for Medical Research. Of course, given Arieti's lack of proficiency in English, he could not work with patients, and therefore was assigned to experimental research (SAPe; SAP 1940h; SAP Um).

Under the direction of Nolan Lewis, the New York Psychiatric Institute hosted many psychiatrists who had escaped from Europe. Among them there were Paul Hoch and Lothar Kalinowski, who conducted the first studies on Electro-Convulsive Therapy in the United States; Zygmunt Piotrowsky, renowned for his studies on the Rorschach Test; and Franz Kallmann, a pioneer of genetic research in psychiatry. Compared to them, the young Silvano experienced a deep feeling of inadequacy and embarrassment. He felt like a "youngster with nothing to offer and much to learn and to receive" (Arieti 1978, p. ix).

Nevertheless, he soon came to know English better. He attended the psychiatric seminars and clinical lectures held at Columbia University, where he met Marie Coleman and Benjamin Nelson, two of his first American acquaintances, with whom, in his still poor and broken English, he tried to hold those conversations that he liked the most, those peppered with "questions for which there is no clear cut answer" (SAP 1974).

In July 1939, Arieti wrote to the Medical Bureau of Chicago, which collected job requests submitted by physicians from all over the country, matching them with openings in the hospital network. Silvano, knowing that his grant was about to expire and that he would soon find himself without funds, tried to act in advance.

The next month he received a disappointing reply:

> My dear Doctor Arieti ... we appreciate the interest you express in our services but from your letter we gain that as yet you have not secured a license to practice in any of the States. Until you have done so, we are doubtful that we shall be able to assist you in making a suitable connection as the clients seeking interns through our Bureau specify that we recommend only graduates of approved Medical Schools. We are enclosing an application which we shall be glad to have you complete and return to us ... If by chance we should learn of some position meeting your requirements in connection with which a graduate of an American school is not required we shall certainly get in touch with you. Meanwhile through your own friends and acquaintances we trust you will be able to make a suitable connection before long (SAP 1939c).

As a matter of fact, Arieti at that time lacked the qualifications and requirements for a medical license, as he had not yet received the validation of his medical degree, which would come from the U.S. Consulate in May 1940 (SAPc; SAPg). Thus, the young man was trapped in a limbo; in America he had to face the obstacles from being an immigrant.

Communication with his family in Italy was also difficult, a circumstance that weighed on him heavily. He wrote to his parents on a regular basis, hoping that his letters were not going to be suppressed by censorship. Only one year after arriving in the United States, he felt tired, as though suddenly aged. His parents in their letters tried to reassure and encourage him not to give up. "Do not think yourself old," wrote his father; "At your age you have to struggle a little in order to settle down. When these troubles are past, know that you will feel younger than you do today. Cheer up, have fun when you can, and take the world as it comes." "I'm sorry that you tell me that you feel old," his mother reiterated. "I ask you fervently to take care of your health. Get some calcium injections and try to eat a few more fresh eggs. Do me a favor, take care of yourself, try to keep well and do not study too hard as you usually do … One day you will be a great professor, an adult man but not at all an old man, and you will find me turned a bit white, but always calm and strong" (SAP 1940b, 1940d, my translation). In these ways Elio and Ines Arieti tried to soothe the castaway's troubles on a foreign island, treating his unexpected condition as an ordinary stage of development.

In January 1940 the Education Department of the University of the State of New York communicated to Arieti that he had passed the examination in English for foreigners (SAP 1940a). Although he was not yet proficient in the language, passing the examination was an important step. His parents rejoiced and congratulated him. "My dearest Silvano, we received with great joy the news that you have passed the English examination … I'm sure that you will have a bright future and will become a true scientist, as you used to write in your letters when you were a child" (SAP Uh, my translation).

Though there was a long road ahead before achieving this goal, Silvano did not slacken his efforts. Knowing that he still had much to learn, in May 1940 he decided to go to New Haven to the Primate Biology Laboratories directed by Robert Yerkes, professor of psychobiology at Yale (SAP 1940h).

Yerkes was a leading figure of American psychology. Already in 1917 he served as President of the American Psychological Association. A master of functionalism, he furthered the cause of a non-reductionist psychobiology, which he conceived as a comprehensive and wide-ranging discipline open to many different issues—from animal behavior, to comparative psychology and human behavior, both individual and collective (Yerkes 1932; Hilgard 1965). It was probably with him that Arieti discovered an important book for his future studies on schizophrenia, Heinz Werner's *Comparative Psychology of Mental Development,* published in 1940 (Werner 1940).

From a methodological point of view, Yerkes's approach was equally wide-ranging. It extended from basic research to applied psychology; from introspection to observation; and from mental testing to laboratory behavioral research in the foot-steps of Ivan Pavlov, whose work Yerkes and Sergius Morgulis had introduced into United States in 1909 (Yerkes & Morgulis 1909). Arieti went to Yale planning a study on the functions of temporal lobes in monkeys, a work to be carried out with Ferraro in New York. To accomplish this plan Arieti needed to learn the methods of behavioral research. In addition, he wished to broaden the study with a comparative methodology; he planned therefore to correlate information obtained in animal experimentation with the structural and functional data available for human infants (SAP 1940h).

Knowing this plan, Yerkes introduced Arieti mindfully to Arnold Gesell, Head of the Yale Clinic of Child Development: "I shall much appreciate it … if you will allow him to come to you for conference. He is very conscious of his linguistic difficulties and is inhibited thereby" (SAP 1940c).

Yerkes, who admired the "natural modesty and shyness" of this "intelligent, earnest, and eager" young man, actually took his situation quite to heart and interceded with the Dazian Foundation for a renewal of his grant: "Starting as he did with a linguistic handicap which embarrassed him …, slow adaptation … was expected. Instead, he has adjusted rapidly and unquestionably has gained much of advantage from his practical experience … and from his professional contacts in Yale. We think very well indeed of his plans of research, and it is a pleasure to be able to report favorably and with enthusiasm on one of the Fellows of the Dazian Foundation and also to support strongly his reappointment" (SAP 1940e, 1940f). At the same time, Yerkes recommended Arieti to Eugene Kahn, Chairman of the Yale Department of

Psychiatry, so that Kahn could offer him some suggestions for a future internship (SAP 1940g).

Beyond Yerkes' affectionate benevolence, Arieti also found in New Haven a stimulating environment and a warm human atmosphere. In those years Yale was a meeting place of many young students coming from abroad, with whom he established enduring contacts and friendships (SAP 1963).[3]

In July 1940, Arieti returned to New York to work on four scientific articles—on histopathological findings in the brain of monkeys subjected to Metrazol shock, experimental encephalitis, the psychic symptoms in pernicious anemia, and general paresis (SAP 1940h). In this research Arieti's neuropathological perspective differed greatly from the "cerebral mythology" that in those years characterized Italian psychiatry's biological approach to the study of the relationship between psychological symptoms and neurological conditions (Babini, Cotti, Minuz & Tagliavini 1982; Guarnieri 1991; Passione 2006, 2007, 2013). Ferraro's work introduced Arieti to a viewpoint different from what he had learned during his university years. In the study on pernicious anemia, for example, there was evidence of cerebral lesions both in subjects with psychiatric complications or psychological symptoms and in those without them. This evidence led Arieti and Ferraro to assume that even in clear-cut neurological and organic cases, brain factors did not have an exclusive role, and they had to be interpreted, instead, as precipitant causes. "Presumably," they wrote, "additional psychogenic and constitutional factors must be at play for the development of mental symptoms" (Ferraro & Arieti 1945, p. 236; Fig. 2.7).

In New York Arieti also continued research he had begun at Yale, for which he needed to acquire monkeys, which the Institute did not have, for the war in Europe prevented the shipping of animals from overseas. Thus, Arieti had to wait a few months until a "conspicuous number of monkeys" were finally delivered to the Institute (SAP 1940i, 1940j). "In the meantime," he wrote to Yerkes, "I should like very much to attend your meetings, as I can readily run up to New Haven on Friday afternoons and profit again by the teachings of your school, if this meets with your approval" (SAP 1940l). Of course, this meant no rest for him on weekends.

Soon, however, Silvano Arieti succeeded in organizing a well-equipped laboratory where monkeys would undergo behavioral training, and where he could carry out delayed response and discrimination experiments (SAP 1941c; Fig. 2.8). He was fully dedicated to this work, treating the animals with care and attention. Sometimes it almost seemed that he wanted to talk with them, as a colleague described Arieti's countertransference-like reaction to Pete, a "mischievous chimp" with an ability to clown around and make fun of him: "I don't believe that monkey likes me," he mused (SAP Un).

The laboratory became a bridge between the big city and New Haven, and Arieti was proud to show it to the colleagues who came to visit from Connecticut. One

[3] In 1963 one of them, Vincent Howles wrote to Arieti: "Dear Silvano, what a delightful surprise to find your note … in my mail today. Since yesterday was my 50th birthday, I have very much been in the mood to remember old friends, three Europeans who made visit to Yerkes' Lab … when I was there – Arieti, Erikson, Tinbergen."

Fig. 2.7 Silvano Arieti at New York Psychiatric Institute. Neuropathological research, *circa* 1940 (Personal Collection of James Arieti) © Courtesy of James Arieti—All rights reserved

Fig. 2.8 Silvano Arieti at New York Psychiatric Institute. Experimental research on monkeys, 1940 (SAP, Library of Congress, Manuscript Division) © Courtesy of James Arieti—All rights reserved

of those who visited the lab was Robert Malmo, a leading figure in the study of the neurological mechanisms of emotion—a subject that was also one of Arieti's main interests (SAP 1941b, 1941e).[4] In his study of emotional changes following brain damage, he observed in animals that had undergone surgery a characteristic syndrome described few years earlier by Paul Bucy and Heinrich Klüver, a pupil of the experimental psychologist Max Wertheimer (Nahm & Pibram 1998). This syndrome manifested a state of insensitivity and extreme apathy. Furthermore, these monkeys showed gross perceptual alterations and discriminative deficiency, such that they could react to a stimulus only disjointedly. Another feature of the syndrome was a habit of bringing anything to their mouths with automatic gestures and swallowing it, regardless its edibility. This was a habit that Arieti would observe one year later also in the schizophrenic inmates of the Pilgrim State Hospital (Arieti 1944a).

After returning from Yale, Arieti carried out some studies of brain tumors, focusing on their vascularization and growth mechanism (Arieti 1942, 1944c; SAP 1948).[5] In this groundbreaking field of medical research he collaborated with Leo Alexander, who a few years later would be called to the Nuremberg Trial as an expert witness (SAP 1941d).

Silvano Arieti's first publication appeared in 1941 in *The American Journal of Psychiatry* (Arieti 1941). It was an important accomplishment for him in view of his search for a job—a search that was becoming more urgent as the expiration of his grant approached (SAP 1941f).[6] By the end of 1940 Silvano Arieti had applied to the Rockefeller Foundation for a fellowship, but, despite Yerkes' and Khan's recommendations, his application was rejected, because the Foundation required a prior one-year psychiatric internship for eligibility, which the young man did not yet have to his credit (SAP 1940k, 1940m).

Thus, finding a hospital for his internship proved to be very difficult despite Arieti's solid scientific background. The main problem was that he could not obtain a medical license because he lacked American citizenship (SAP Ua). The only way to take a medical licensing examination *before* becoming an American citizen would have been to obtain a regular "alien registration card" from the Immigration and Naturalization Service, for which Arieti put in a request. After a long wait, he received the card in February 1942 (SAP 1940n, 1940o, 1942b).

Meanwhile, many young colleagues whom he had met during his early work experiences were finding positions around the country. One of these wrote from Orange Park, Florida: "There are many affairs of yours that I should enjoy hearing about: your experiments on behavior, your progress toward a medical license, your plans for the immediate future. Your feelings about the European situation must be painfully conflicting now, because of your sympathy both with international democracy and

[4] Not by chance, in January 1941 Arieti wrote to Papez to ask him for a copy of his article on the mechanism of emotion, published in 1937.

[5] In 1948 the American Cancer Society listed Arieti's contributions in a series of outstanding articles on cancer.

[6] Expiration date was November 1941.

with the well-being of friends at home" (SAP 1941a). And one cannot in fact doubt that Arieti's mind was consumed with worries about his family and friends in Europe.

In the summer of 1941, Robert Yerkes, who had truly taken to heart Arieti's difficult situation, wrote to some colleagues to advance the cause of this willing young man (SAP 1941i). Yerkes soon received a response from Arthur Ruggles, the Superintendent of the Butler Hospital in Providence, Rhode Island. Without wasting time, Yerkes asked his secretary, Helen Marford, to write to Arieti:

> Dear Doctor Arieti, Doctor Ruggles has suggested that you write him concerning the possibility of openings in psychiatric institutions. He already has a copy of the curriculum vitae which you sent Doctor Yerkes last week. In his letter Doctor Ruggles mentioned also a vacancy at the Pilgrim State Hospital in New York. His address is: Doctor Arthur H. Ruggles, Butler Hospital, Providence, Rhode Island. Doctor Yerkes left this morning for Franklin, New Hampshire (SAP 1941h).

Thus "the largest psychiatric hospital in the world" (Arieti 1978, p. ix) entered Arieti's life. A few days after the above-mentioned letter Arieti went to Long Island for a visit. Formally, he had to examine a patient there, on Ferraro's behalf (SAP 1941j). Unofficially, and more likely, he also wanted to take a look at the locale that might become his future workplace.

The first impression was not very exciting, and Arieti went to Providence, in August 1941, with the great hope of obtaining an alternative position at the Butler Hospital, where he knew that there was also an opening. Raising his hopes, Yerkes and Ferraro recommended him for the job in Providence. "I am sure you will find him a capable young man whose character and integrity I can recommend and who, I am certain, will be an asset for your Institution," Ferraro wrote to Ruggles (SAP 1941k).

Silvano Arieti awaited eagerly and with confidence the result of his job interview, but things did not turn out as he had hoped. In spite of his scientific qualifications, Ruggles felt that Arieti's language handicap, "while not great," was such that it was better to choose another candidate. "As I indicated to you when you were here," Ruggles remarked, "such a handicap has proved disturbing to a number of our patients" (SAP 1941l).

With these words a tired and dejected Silvano reported what happened to Yerkes:

> Dear Dr. Yerkes ... I am writing to you in relation to my conference with Dr. Ruggles. A few days after his return from vacation, Dr. Ruggles interviewed me in Providence and very kindly welcomed me. He mentioned your recommendation and had many words of encouragement for me ... On Tuesday, September 2nd, I received a letter from Dr. Ruggles, in which he notified me that he did not consider it convenient to appoint me because even a slight linguistic handicap is not well accepted by private patients. Accordingly, as you can well imagine, Dr. Yerkes, this experience left me quite discouraged (SAP 1941m).

Rejected by Providence, Silvano Arieti was accepted by the Pilgrim State Hospital—a large State Mental Hospital where his linguistic handicap did not matter much, since words were not needed for the merely custodial work that was expected of him there.

The young doctor entered Pilgrim State Hospital on November 1, 1941, as an intern at a yearly salary of 1.800 dollars (SAP 1941n). He remained there until 1946, five difficult years, and yet, amazingly rich ones.

2.3 On the Island

Pilgrim State Hospital, Brentwood, NY, 1944. On the last page of his notebook Silvano Arieti signs his name at the end of a "three-acts play" he has written during breaks from work, a drama entitled *I Naufraghi* (*The Castaways*). The title is suggestive since he himself was a castaway, having been hit by a storm that suddenly changed the course of his life. After the ocean crossing that took him from his home, he had to navigate his own course on the American shore in order to arrive at an even more faraway harbor—a "blessed island" that offered a sanctuary "where hope [was] still allowed" while, at the same time, a "damned island" because of its desolation—a place where even the "slightest intellectual ambitions" seemed to have no room (SAP 1944d, my translation).[7]

These are the two opposite points of view expressed by the characters in Arieti's play, the setting of which is reminiscent of the well-known Platonic myth of the cave. In *I Naufraghi,* castaways take shelter in a cavern that is illuminated by a ray of light filtering in from the outside, and the sunken ship's captain urges his fellows to follow the light, inviting them to go beyond the evident presence of their misfortune and to broaden their view, so "that not even this corner of the globe be stolen from a human domain and knowledge" (SAP 1944d, my translation).

The metaphor of the island describes how Arieti probably experienced taking up residence at Pilgrim State Hospital. It was a "blessed island," since, in the face of adversity, he had finally found a job; but it was also a "damned" island, for it was the largest psychiatric hospital in the world, the place where he came "face to face with the ravages of mental illness" (Bemporad 1981, p. iv). It was a gloomy and squalid stage that in its overwhelming overcrowdedness made any social life difficult, even with the colleagues with whom Arieti shared the staff apartments. In this environment there was little to be glad about, especially for someone like him, whose nature was shy and withdrawn. Assigned to a mainly custodial and woefully underpaid work, Silvano Arieti endured his early days at Pilgrim State Hospital with Stoic resolve (SAP Un). His situation was made worse by his anguish over the fate of his Italian loved ones, with whom any correspondence became slower and rarer as the war progressed (SAPa).

Nevertheless, he did not succumb to homesickness and the sorrowful separation from his roots; instead, he persistently tried to find his way in his new land. Newly appointed to Pilgrim State Hospital, he married Jane Jaffe at the Spanish and Portuguese Synagogue, in New York, on December 14, 1941 (the day before his

[7] This manuscript bears in the Archives the indication of place (Brentwood) and date (1944), but has been erroneously catalogued as "undated".

parents' wedding anniversary) (SAPh). Thus, in this difficult time Silvano Arieti did not abandon himself to despair. He stayed the course. He seemed almost to be carrying the paradoxical unity of precariousness and solidity embodied in the Leaning Tower of Pisa, which had always fascinated him (Arieti 1979a).

He knew, of course, that he had a long way to go. To obtain a medical license he had to serve a one-year internship, and then he had to qualify as a psychiatrist. It was a demanding agenda, and Arieti worked hard full time. Following the words of his play's imaginary sunken ship's captain, he saw Pilgrim State Hospital as a faraway "corner of the globe" that could be redeemed by human knowledge.

Despite its stultifying daily routine, the largest mental hospital in the world abounded in opportunities to learn about schizophrenic patients, and Arieti, who had been assigned to their surveillance, lost no chance to learn from his surroundings. He filled notebooks with ideas and questions about the thought processes and feelings of the patients. "Abstract ideas. Their functioning in schizophrenia. To be investigated." And also, "Schizophrenia: is really losing emotional skills a basic characteristic, or does the loss of such skills result from the interruption of associative processes?" (SAP Ud, my translation).

Still lacking in an education about dynamic processes, he tried to answer these questions by turning to the work of Eugen Bleuler—his first resource in approaching schizophrenia (Bleuler 1924). Finding answers to questions about schizophrenia was not an easy task, though, since patients in advanced stages of this illness speak a language that psychiatric treatises used to describe as a meaningless "word-salad." For Arieti, whose English still lacked proficiency, understanding the utterances of the patients at Pilgrim State Hospital was a big obstacle to overcome; yet, it was by listening to a schizophrenic's incoherent chatter that little by little he succeeded in mastering the American language and settling into his "stepmother" tongue (Passione 2018).

Therefore, despite the mere surveillance work required of him, Arieti accomplished much more at Pilgrim State Hospital. He listened, studied, observed, questioned, and thought a lot about what he saw. He was lucky enough to have a supervisor like Newton Bigelow, who recognized his desire to do research and encouraged him to use the large number of clinical cases at the mental hospital as an opportunity for scientific growth (SAP 1972).[8]

Arieti promptly followed his suggestion, focusing on a problem that had especially interested him since his early days at the New York Psychiatric Institute—the relationship between psychological symptoms and neurological conditions. In fact, he devoted his first article as an intern at Pilgrim State Hospital to this subject, an article published, with Bigelow's help, in 1943. The paper was the case report of a

[8] About Bigelow Arieti wrote: "Dr. Bigelow offered precious help. He realized that I had been in this country only for a short time. I could barely communicate in broken English. I had very few friends and I was unfamiliar with the American ways. In spite of all these shortcomings, Dr. Bigelow recognized in me a real concern for the mentally ill and the desire to do research, and he helped me as much as he could".

forty-six-year-old woman admitted to the hospital in 1942 with a diagnosis of functional psychosis, of which Arieti was doubtful. He thought she had a brain tumor, a suspicion that would later be confirmed by an autopsy (Arieti 1943).

Compared to Italian psychiatry, which at that time lacked any psychological investigations and was characterized mainly by an organic approach, his article was very different. In the case of this patient, Arieti's suspicion of a brain tumor did not stem from the neurological examination of the subject, which, in fact, was essentially negative; it stemmed from a clinical and psychological exam, where Arieti, remembering Piotrowsky's lesson at the New York Psychiatric Institute, had given a Rorschach Test. In short, Arieti was able to question the original diagnosis by means of an observation that focused on a psychological evaluation.

In November 1942, Silvano Arieti finally received his medical license (SAP 1942a). The law stated that the license had to be registered in the Clerk's Office of the county where the titleholder intended to practice. For this reason, Arieti enrolled at the Register of Physicians and Surgeons of the Bronx, where in the meantime he had transferred his residence (SAPe; SAPg).

After receiving a medical license, Arieti broadened the range of his practice. Starting in 1943, he served as assistant neurologist in the Vanderbilt Clinic of Presbyterian Hospital. In the same year he became a member of the American Psychiatric Association and of the American Medical Association (SAP 1943c, 1943e, 1946). He also expanded his collaboration with Armando Ferraro in a study of the neuropsychiatric implications in tropical diseases (SAP 1943g, 1943i), and attended the meetings of the American Association of Neuropathologists, where he met Heinrich Klüver (SAP 1942c, 1943j).[9]

Still, he continued to be based at Pilgrim State Hospital, where he continued his internship, working zealously despite a general "climate of therapeutic passivity and resignation" (Arieti 1973b, p. 334). In this regard, we should keep in mind that at that time the psychopharmacological revolution had not yet arrived, and that, moreover, Pilgrim State was not a mental hospital where psychotherapy was practiced systematically, as it was, for example, at the Chestnut Lodge in Rockville, Maryland. Nevertheless, Arieti did not lose heart. With his patients—even the most regressed ones—he tried everything he could to improve their condition. Though he still lacked a proper training in dynamic psychiatry and psychoanalysis, he tried to penetrate the psychic complexes of his patients, whom he encouraged to draw, and then questioned their drawings. He paid attention to their word-salad. In short, he did not subscribe to Bleuler's theory of schizophrenic withdrawal but always tried to reach patients, to establish a relationship with them—the only approach that seemed effective, even in the ostensibly most hopeless cases (Arieti 1973b; SAP Ue).

Patients recognized Arieti's concern for them. One, for example, continued to write to him after being discharged from the hospital, describing his life "outside"

[9] In 1942 Arieti presented a paper entitled "The Vascularization of Cerebral Neoplasms Studied with the Fuchsin Stain." The next year he presented two papers: the first one, with Ferraro and Roizin, entitled "Cerebral Pathology in Experimental Vitamin K Deficiency" and the second one (by him only), entitled "Multiple Meningioma and Meningiomas Associated with Other Brain Tumors".

and pleading with this young doctor not to work too hard (SAP 1943a). In this way, Arieti spent his days in the largest psychiatric hospital in the world, where all human feelings mingled: desolation, resignation, suffering, warmth, friendship, and trust.

In the meantime, his application for citizenship was pushed forward (SAPi). Here, again, Bigelow offered help with the Immigration and Naturalization Service as a guarantor: "I have known Dr. Arieti for about one year as a member of the medical staff of this hospital. He has worked under my direction and I have been in conversation with him several times ... From his statements and behavior, it is my opinion that he is loyal to the United States of America" (SAP 1943b). The procedure was lengthy, but it finally concluded on August 31, 1943, when Silvano Arieti finally became an American citizen (SAPg; SAPi).

In the spring of the following year, he appeared before the American Board of Psychiatry and Neurology (Aminoff & Faulkner 2012) to qualify officially as a specialist. He was nervous and stressed, but found the commissioner kind and encouraging, and everything went well (SAP 1953). A few days after the exam, the Board's secretary, Walter Freeman, communicated to him that he had been awarded the diploma in psychiatry and neurology (SAP 1944b).

Meanwhile, his work with schizophrenics began to bear fruit, as Arieti's first paper on schizophrenia appeared in 1944 in *Psychiatric Quarterly*. Entitled "An Interpretation of the Divergent Outcome of Schizophrenia in Identical Twins," the article discussed the clinical history of two monozygotic twin sisters, Magda and Selena, both admitted to Pilgrim in 1942 (Arieti 1944b).

In the 1940s, the genetics of schizophrenia was a cutting-edge field of research, the importance of which had been emphasized just a few years before by Nolan Lewis in *The Year Book of Neurology, Psychiatry and Endocrinology*. The New York Psychiatric Institute, which Lewis directed, had played a crucial role in this field of research, given an impetus by the work of Franz Kallmann (Kallmann 1938; Kallmann & Barrera 1942). In his contribution Arieti resumed a line of research that he already had come across in his early years in New York. At the same time, he moved the work forward by advocating a non-unilateral interpretation of "the complex problem of nature-nurture in the pathogenesis of mental illnesses" (Arieti 1944b, p. 587). In the article, the divergent outcomes (recovery in one case, slight improvements in the other) of two identical twins brought up in the same environment showed, according to Arieti, the unsuitability of any etiological unilateralism: both merely genetic and merely environmental interpretations of schizophrenia were unsatisfactory. Between these opposite extremes, Arieti suggested a third point of view, one that considered both the role played by the two women's different psychological attitudes *and* the biological mechanisms studied by Kallmann.

Hence, Arieti's first paper of schizophrenia already showed features that characterize his later thought on the illness: the call for *interpretation* in psychiatric work (the word "interpretation" appears for the first time in the title of this paper); the call for integrated and complementary views beyond one-sidedness; and the role played by characteristic psychological features of the individual.

This approach developed in his next paper, which appeared in *The Journal of Nervous and Mental Disease* under the title "The 'Placing-Into-Mouth' and

Coprophagic Habits Studied From a Point of View of Comparative Developmental Psychology." In this article, Arieti took into account the schizophrenic's habit of "picking up from the floor and putting into their mouth inedible objects found in their immediate environment and within easy reach," a phenomenon he regularly observed at Pilgrim State Hospital, where on the autopsy table it was a relatively common experience "to find in the stomach or intestines of patients who were affected by the most advanced stages of dementia praecox, spoons, stones, pieces of scrap iron, wood, paper, cores, etc." (Arieti 1944a, p. 959).

This particular symptom of advanced schizophrenia had already been described by Bleuler, who noticed that it was not rare to see patients grasp their own feces and eat them. Nevertheless, restricting himself to the simple description of this habit, Bleuler did not question its meaning, interpreting it as an expression of a demented schizophrenic's silly and purposeless behavior. Arieti, on the contrary, thought that this symptom was impelled by deeper causes, and he interpreted it as the expression of mechanisms belonging to lower levels of nervous integration, i.e., to more primitive stages of development.

In coming up with this interpretation, which represents the first step toward the idea of schizophrenia as a state of regression to primitive levels of development, Arieti remembered a lesson he had learned at Yale, for the grasping and coprophagic habits that he observed in Pilgrim State Hospital's inmates reminded him of what Klüver and Bucy had described in their studies of monkeys subjected to brain surgery.

Using the developmental frame of reference provided by Heinz Werner's work, Arieti correlated animal findings with the available human evidences coming from different disciplines—Kurt Goldstein's neurological studies describing such habits in brain-injured patients; Leo Kanner's child psychiatry describing them in one-year old children; the anthropological observations of geophagy in savage tribes; and the psychoanalytical description of an oral stage of development. We have here another example of the integrative and complementary approach that had been at the center of Arieti's study of identical twins: "it seems, therefore," he wrote, "that we have here a typical illustration of the inter-relations … of psychiatry, psychoanalysis, and neurology, even in its experimental part" (Arieti 1944a, p. 963). It was an approach put forth also by Roy Grinker, to whom Arieti referred in the article (Grinker 1941; SAP 1941g).

Through the analysis of a particular feature of schizophrenic symptomatology, Arieti revamped an idea formulated in 1924 by Alfred Storch, that schizophrenia is a form of regression to archaic and primitive functions. More precisely, the coprophagic habit could be seen as a regression to a primitive way of reacting, in which the regulative and inhibiting action of the nervous system's higher levels was not yet available. "In other words, we may deal with one of those responses in which a short-circuiting takes place between the functions of reception and those of reaction instead of the usual way with participation of the higher centers. These reactions are intermediate between reflexes and voluntary acts, having some characteristics of compulsory acts" (Arieti 1944a, p. 963).

Arieti's contribution did not go unnoticed. In the opinion of Norman Cameron, Chairman of the Department of Psychology of the University of Wisconsin, its most interesting trait was "the factual way" in which Arieti handled the material, as well as his attempt to bring together different observations taken from the field of child psychiatry, anthropology, comparative behavior, neurology, and psychoanalysis. "I also like your restraint in not going off into long, positivistic theories," Cameron wrote to him. "I hope you do more reporting of this kind; and when you do, don't forget my address" (SAP 1944c).

Cameron's comments did not fall on deaf ears. After "The 'Placing-Into-Mouth' and Coprophagic Habits," Silvano Arieti continued his detailed analysis of schizophrenic symptoms in the various stages of the illness. In his next paper, for example, he focused on the terminal stage of schizophrenia and analyzed its main regressive features (Arieti 1945a). In this study Arieti openly declared his debt to Kurt Goldstein, to whom he owed both the concept of regression as an explanatory key of schizophrenia, and the emphasis on the role played by psychological factors on functional changes of the nervous system (Goldstein 1943).

Shortly afterwards, he broadened his analysis by taking into account the preterminal stage of schizophrenia, describing its typical "hoarding" and self-decorating habits (Arieti 1945b). Here he further discussed the link with psychoanalysis (which he had briefly mentioned in other papers) with references to Freud's, Abraham's and Ferenczi's observations on the hoarding habit (Freud 1938; Ferenczi & Jones 1916; Abraham 1927), whereas the link with anthropology was strengthened by means of the appearance, in the bibliographic references, of Boas' writings on primitive art (Boas 1927).

Thus, on the one hand Arieti's years at the Pilgrim State Hospital were rich with scientific achievements. In 1944, he was also appointed at Columbia University as an unsalaried research assistant in psychiatry (SAP 1944a). On the other hand, they were hard personally. In Europe, the war had reached a dramatic peak, and he anguished over his dear ones in Italy. With great disquietude, he read the news about the war in American newspapers every day. And to make matters worse, his region of origin was often prominently in the news, since Pisa was on the Gothic Line. His correspondence with his parents mirrors the dramatic situation. On November 30, 1942, at 11 p.m., during the hardest days for the Jewish people in Pisa, Elio and Ines Arieti wrote a letter to their sons in America:

> Our dearest Silvano and Giulio, we have now reached the hardest and most dangerous time in our existence, when we have to try by any means to save and preserve ourselves for you. May this sheet of paper, written in such sad hours, bring you our affection and our dearest feelings of fatherly and motherly love. We are to blame ... for letting you leave alone ... We knew you were strong and intelligent, and various considerations, including that of not being in the way ..., led us to do so. Be strong and do accept what God in his great justice wanted for us. This is our moral testament: always be good and honest, and always love humble and needy persons. They are the ones closest to God. Always love freedom, with all your strength, and abhor tyranny. Always love one another, thinking that we love you very much, and that we will always love you (SAP 1943h, my translation).

Just few days before the writing of this letter, Silvano Arieti had contacted the War Department, asking to serve in the armed forces. He wanted to enlist and to contribute to the war effort; above all, he wanted to defeat the sense of helplessness and urgency he felt every day as he anxiously rushed to read the newspapers (SAPb). His application, however, was rejected (SAP 1943f), because during the war the U.S. government had established a "freezing" rule that required public officers to stay at their posts for the duration of the war (Arieti 1979b; SAP 1979a, 1979b; SAP Un).

Although now, as an American citizen, he was eligible for the draft and had all the professional qualifications required to leave Pilgrim State Hospital, Arieti remained *stuck on the island*. He could not join the armed forces, nor could he change jobs. Despite the help of Newton Bigelow, who moved to Albany and would gladly have brought Arieti with him, he remained "frozen" at his workplace. After all, as Carl Warden wrote from the Department of Psychology of Columbia University, "the war comes first" (SAP 1943d; Minnick, Warden & Arieti 1946). Better times were yet to come.

References

Abraham, K. (1927). The Psychosexual Differences Between Hysteria and Dementia Praecox. In K. Abraham & E. Jones (Eds.), *Selected Papers on Psychoanalysis* (pp. 64–79). Hogarth Press.
Alger, I. (1983). The Intellect and Humanism of Silvano Arieti. *The Journal of the American Academy of Psychoanalysis,* 11 (1), 15–34.
Aminoff, M. J., Faulkner, L. (2012). *The American Board of Psychiatry and Neurology: Looking Back and Moving Ahead.* American Psychiatry Publishing.
Arieti, J. (1999). Memories of the Son of a Psychiatrist. *Journal of the American Academy of Psychoanalysis,* 27 (4), 541–550.
Arieti, S. (1941). Histopathological Changes in Experimental Metrazol Convulsions in Monkey. *The American Journal of Psychiatry,* 98 (1), 70–76.
Arieti, S. (1942). The Vascularization of Cerebral Neoplasm Studied with the Fuchsin Method of Eros. *The Journal of Neuropathology and Experimental Neurology,* 1 (4), 375–393.
Arieti, S. (1943). Frontal Lobe Tumor Expanding Into the Ventricle. Clinicopathologic Report. *Psychiatric Quarterly,* 17 (2), 227–240.
Arieti, S. (1944a). The "Placing-Into-Mouth" and Coprophagic Habits Studied From the Point of View of Comparative Developmental Psychology. *The Journal of Nervous and Mental Disease,* 99 (6), 959–964.
Arieti, S. (1944b). An Interpretation of the Divergent Outcome of Schizophrenia in Identical Twins. *Psychiatric Quarterly,* 18 (4), 587–599.
Arieti, S. (1944c). Multiple Meningioma and Meningiomas Associated with Other Brain Tumors. *The Journal of Neuropathology and Experimental Neurology,* 3 (3), 255–270.
Arieti, S. (1945a). Primitive Habits and Perceptual Alterations in the Terminal Stage of Schizophrenia. *Archives of Neurology and Psychiatry,* 53 (5), 378–384.
Arieti, S. (1945b). Primitive Habits in The Preterminal Stage of Schizophrenia. *The Journal of Nervous and Mental Disease,* 102 (4), 367–375.
Arieti, S. (1959). Schizophrenic Thought. *The American Journal of Psychotherapy,* 13 (3), 537–552.
Arieti, S. (1968). Some Memories and Personal Views. *Contemporary Psychoanalysis,* 5 (1), 85–89.
Arieti, S. (1972). *The Will To Be Human.* Quadrangle Books.
Arieti, S. (1973a). Anxiety and Beyond in Schizophrenia and Psychotic Depression. *The American Journal of Psychotherapy,* 17 (3) 338–345.
Arieti, S. (1973b). Schizophrenic Art and its Relationship to Modern Art. *The Journal of the American Academy of Psychoanalysis,* 1 (4), 333–365.
Arieti, S. (1974). *Interpretation of Schizophrenia.* Basic Books.

Arieti, S. (1976). *Creativity: The Magic Synthesis.* Basic Books.
Arieti, S. (1978). *On Schizophrenia, Phobias, Depression, Psychotherapy and the Farther Shores of Psychiatry.* Brunner Mazel.
Arieti, S. (1979a). *The Parnas.* Basic Books.
Arieti, S. (1979b). From Schizophrenia to Creativity. *The American Journal of Psychotherapy,* 33 (4), 490–505.
Arieti, S. (2001). Gli anni pisani di Silvano Arieti. In R. Bruschi (Ed.), *Uno psichiatra tra due culture, Silvano Arieti 1914–1981* (pp. 81–90). ETS.
Babini, V. P. (2009). *Liberi tutti. Manicomi e psichiatri in Italia: una storia del Novecento.* Il Mulino.
Babini, V. P. (2014). Looking Back: Italian Psychiatry From Its Origins to Law 180 of 1978. *The Journal of Nervous and Mental Disease,* 202 (6), 428–431.
Babini, V. P., Cotti, M., Minuz, F., & Tagliavini, A. (1982). *Tra sapere e potere. La psichiatria italiana nella seconda metà dell'Ottocento.* Il Mulino.
Bateman-House, A., Fairchild, A. (2008). Medical Examination of Immigrants at Ellis Island. *American Medical Association Journal of Ethics,* 10 (4), 235–241.
Bemporad, J. (1981). In Memoriam. Silvano Arieti 1914–1981. *The Journal of the American Academy of Psychoanalysis,* 9 (4), iii–vii.
Bleuler, E. (1924). *Textbook of Psychiatry.* Macmillan Company.
Boas, F. (1927). *Primitive Art.* Aschehoug & Company.
Cabanis, P. J. G. (1798). *Du Degré de Certitude de la Médecine.* Crapelet.
Clemmens, E., Spiegel, R., Bieber, I., & Di Cori, F. (1982). Silvano Arieti: 1914–1981. *Academy Forum,* 26 (1) 6–9.
Ferenczi, S., Jones, E. (1916). *Contributions to Psychoanalysis.* Richard G. Badger.
Ferraro, A., & Arieti, S. (1945). Cerebral Changes in the Course of Pernicious Anemia and Their Relationship to Psychic Symptoms. *The Journal of Neuropathology and Experimental Neurology,* 4 (3), 217–239.
Freud, S. (1938). *The Basic Writings of Sigmund Freud.* Edited by A. A. Brill. Modern Library.
Goldstein, K. (1943). The Significance of Psychological Research in Schizophrenia. *The Journal of Nervous and Mental Disease,* 97 (3), 261–279.
Grinker, R. (1941). The Interrelations of Neurology, Psychiatry and Psychoanalysis. *The Journal of the American Medical Association,* 116 (20), 2236–2241.
Guarnieri, P. (1991). *La storia della psichiatria. Un secolo di studi in Italia.* Olschki.
HAUP. *Silvano Arieti. Personal File.*
Hilgard, E. (1965). *Robert Mearns Yerkes 1876–1956. A Biographical Memoir.* National Academy of Sciences.
Kallmann, F. (1938). *The Genetics of Schizophrenia.* Augustin.
Kallmann, F., Barrera, E. S. (1942). The Heredo-Constitutional Mechanisms of Predisposition and Resistance to Schizophrenia. *The American Journal of Psychiatry,* 98 (4), 544–550.
Minnick, R. S., Warden, C., & Arieti, S. (1946). The Effects of Sex Hormones on the Copulatory Behavior of Senile White Rats. *Science,* 103 (2687), 749–750.
Moravia, S. (1974). Medicina ed epistemologia nel giovane Cabanis. In P.J.G. Cabanis, *La certezza della medicina* (pp. VIII–XXXVIV). Laterza.
Nahm, F., Pibram, K. (1998). *Heinrich Klüver 1897–1979. A Biographical Memoir.* National Academy Press.
Passione, R. (2006). *Ugo Cerletti. Scritti sull'elettroshock.* Franco Angeli.
Passione, R. (2007). *Ugo Cerletti. Il romanzo dell'elettroshock.* Aliberti.
Passione, R. (2013). *Per un'epistemologia della complessità. Gaetano Perusini nella storia della psichiatria italiana.* Aracne.
Passione, R. (2018). Language and Psychiatry: The Contribution of Silvano Arieti Between Biography and Cultural History. *European Yearbook of the History of Psychology,* 4, 11–36.
Passione, R. (2021). Silvano Arieti (1914–1981). In P. Guarnieri (Ed.), *Intellectuals Displaced from fascist Italy. Migrants, Exiles and Refugees Fleeing for Political and Racial Reasons.* Firenze University Press. http://intellettualinfuga.fupress.com/en

Roizin, L. (1983). In Memoriam. Armando Ferraro, M.D. (1896–1982). *The Journal of Neuropathology and Experimental Neurology,* 42 (2), 213–215.
Roth, P. (1967). *Portnoy's Complaint.* Random House.
SAPa. *Correspondence, 1940–1981.*
SAPb. *Subject File, 1914–1981. Collected Material. Newspaper Clippings, 1944–1981.*
SAPc. *Subject File, 1914–1981. Education. University of Pisa, Pisa, Italy, 1938–1940, 1965.*
SAPd. *Subject File, 1914–1981. Education. Workbooks, circa 1930.*
SAPe. *Subject File, 1914–1981. Employment. Position Appointments, 1941–1961.*
SAPf. *Subject File, 1914–1981. Personal. Arieti, Elio, 1914–1969.*
SAPg. *Subject File, 1914–1981. Personal. Certificates, 1940–1975.*
SAPh. *Subject File, 1914–1981. Personal. Memento, 1955–1969.*
SAPi. *Subject File, 1914–1981. Personal. Naturalization Documents, 1940–1943.*
SAPj. *Subject File, 1914–1981. Personal. Photographs, circa 1929–1976.*
SAP. (1888). PD (Estratto del registro degli Atti di nascita dell'anno 1888) . In *Subject File, 1914–1981. Personal. Arieti, Elio, 1914–1969.*
SAP (1930). M (Silvano Arieti, *La prima conquista. Dramma in quattro atti*). In *Subject File, 1914–1981. Education. Workbooks, circa 1930.*
SAP (1938a). PD (Libretto del tirocinio pratico in medicina e chirurgia del Sig. Arieti Silvano). In *Subject File, 1914–1981. Education. University of Pisa, Pisa, Italy 1938a–1940, 1965.*
SAP (1938b). C, September 15 (Ayala's reference letter). In *Subject File, 1914–1981. Education. University of Pisa, Pisa, Italy 1938b–1940, 1965.*
SAP (1938c). PD, 1938c (passport of Silvano Arieti). In *Subject File, 1914–1981. Personal. Family Passports, 1938–1968.*
SAP (1939a). C, March 4 (Hans Steck's certification of Arieti's attendance of the University Psychiatric Clinic in Lausanne). In *Subject File, 1914–1981. Education. University of Pisa, Pisa, Italy 1938–1940, 1965.*
SAP (1939b). B, April 1 (Queen Mary. List of Tourist Passengers). In *Subject File, 1914–1981. Personal. Mementos.*
SAP (1939c). L, August 4 (Burneice Larson to Silvano Arieti). In *Correspondence, 1940–1981.*
SAP (1940a). L, January 19 (L. Field to Silvano Arieti). In *Subject File, 1914–1981. Personal. Certificates, 1940a–1975.*
SAP (1940b). L, May 17 (Elio and Ines Arieti to Silvano and Giulio Arieti). In *Correspondence, 1940b–1981.*
SAP (1940c). L, May 23 (Robert M. Yerkes to Arnold Gesell). In *Correspondence, 1940c–1981.*
SAP (1940d). L, June 10 (Elio and Ines Arieti to Silvano Arieti). In *Correspondence, 1940d–1981.*
SAP (1940e). L, June 26 (Robert M. Yerkes to Robert A. Lambert). In *Correspondence, 1940e–1981.*
SAP (1940f). L, June 27 (Robert M. Yerkes to Emanuel Libman). In *Correspondence, 1940f–1981.*
SAP (1940g). L, July 3 (Robert M.Yerkes to Eugene Kahn). In *Correspondence, 1940g–1981.*
SAP (1940h). T, July 5 (Curriculum Vitae of Silvano Arieti). In *Subject File, 1914–1981. Employment. Curriculum Vitae, 1940h–1980 circa.*
SAP (1940i). L, July 12 (Silvano Arieti to Robert M. Yerkes). In *Correspondence, 1940i–1981.*
SAP (1940j). L, July 16 (Walter Grether to Silvano Arieti). In *Correspondence, 1940j–1981.*
SAP (1940k). L, October 7 (Silvano Arieti to Eugene Kahn). In *Correspondence, 1940k–1981.*
SAP (1940l). L, October 26. (Silvano Arieti to Robert M. Yerkes). In *Correspondence, 1940l–1981.*
SAP (1940m). L, November 18. (Silvano Arieti to Eugene Khan). In *Correspondence, 1940m–1981.*
SAP (1940n). L, December 19 (Silvano Arieti to Alien Registration Division, Immigration and Naturalization Service). In *Subject File, 1914–1981. Personal. Naturalization Documents, 1940n–1943.*
SAP (1940o). L, December 27 (Donald Perry to Silvano Arieti). In *Subject File, 1914–1981. Personal. Naturalization Documents, 1940o–1943.*
SAP (1941a). L, January 19. (Vincent Howles to Silvano Arieti). In *Correspondence, 1940–1981.*
SAP (1941b). L, January 27 (Silvano Arieti to James W. Papez). In *Correspondence, 1940–1981.*
SAP (1941c). L, February 28 (Silvano Arieti to Emanuel Libman). In *Correspondence, 1940–1981.*

SAP (1941d). L, March 17 (Silvano Arieti to Leo Alexander). In *Correspondence, 1940–1981.*
SAP (1941e). L, March 18 (Silvano Arieti to Robert B. Malmo). In *Correspondence, 1940–1981.*
SAP (1941f). L, May 5 (The Dazian Foundation to Silvano Arieti). In *Correspondence, 1940–1981.*
SAP (1941g). L, June 9 (Silvano Arieti to Roy Grinker). In *Correspondence, 1940–1981.*
SAP (1941h). L, July 11 (Helen Marford to Silvano Arieti). In *Correspondence, 1940–1981.*
SAP (1941i). L, July 11. (Silvano Arieti to Robert M. Yerkes). In *Correspondence, 1940–1981.*
SAP (1941j). L, July 25 (Silvano Arieti to Armando Ferraro). In *Correspondence, 1940–1981.*
SAP (1941k). L, August 27 (Armando Ferraro to Arthur Ruggles). In *Correspondence, 1940–1981.*
SAP (1941l). L, August 29 (Arthur Ruggles to Silvano Arieti). In *Correspondence, 1940–1981.*
SAP (1941m). L, September 4 (Silvano Arieti to Robert M. Yerkes). In *Correspondence, 1940–1981.*
SAP (1941n). L, October 3 (Harry J. Worthing to Silvano Arieti). In *Subject File, 1914–1981. Employment. Position appointments, 1941n–1961.*
SAP (1942a). C (License to Practice Medicine and Surgery in the State of New York n° 40955). In *Subject File, 1914–1981. Personal. Certificates, 1940–1975.*
SAP (1942b). C (United States Department of Justice. Alien Registration. Certificate of Identification). In *Subject File, 1914–1981. Personal. Naturalization Documents, 1940–1943.*
SAP (1942c). B (American Association of Neuropathologists. Program of the Annual Meeting). In *Subject File 1914–1981. Conferences and Lectures.*
SAP (1943a). L, January 7 (L.S. to Silvano Arieti). In *Correspondence, 1940–1981.*
SAP (1943b). L, April 24 (Newton Bigelow to The Immigration and Naturalization Service). In *Subject File, 1914–1981. Personal. Naturalization Documents, 1940–1943b.*
SAP (1943c). L, June 1 (Austin Davies to Silvano Arieti). In *Subject File, 1914–1981. Employment. Position appointments, 1941–1961.*
SAP (1943d). L, September 23 (Carl J. Warden to Silvano Arieti). In *Correspondence, 1940–1981.*
SAP (1943e). L, November 23 (The American Medical Association to Silvano Arieti). In *Subject File, 1914–1981. Employment. Position appointments, 1941–1961.*
SAP (1943f). L, November 23 (The War Department, Army Service Forces to Silvano Arieti). In *Subject File, 1914–1981. Employment. Position Appointments, 1941–1961.*
SAP (1943g). L, November 26 (Silvano Arieti to Alfred R. Crawford). In *Subject File, 1914–1981. Research. Malaria research grant, 1942–1944.*
SAP (1943h). L, November 30 (Elio and Ines Arieti to Silvano and Giulio Arieti).
SAP (1943i). L, December 31 (Jean Curran to Silvano Arieti). In *Subject File, 1914–1981. Research. Malaria research grant, 1942–1944.*
SAP (1943j) B (American Association of Neuropathologists. Program of the Annual Meeting). In *Subject File 1914–1981. Conferences and Lectures.*
SAP (1944a). L, May 1 (Philip Hayden to Silvano Arieti). In *Subject File, 1914–1981. Employment. Position Appointments, 1941–1961.*
SAP (1944b). L, May 16 (Walter Freeman to Silvano Arieti). In *Subject File, 1914–1981. Employment. Position appointments, 1941–1961.*
SAP (1944c). L, November 28 (Norman Cameron to Silvano Arieti). In *Correspondence, 1940–1981.*
SAP (1944d). M (Silvano Arieti, *I Naufraghi. Dramma in tre atti*). In *Speeches and Writings, 1940–1981. Plays. I Naufraghi. Undated.*
SAP (1946). L, September 6 (Edwin G. Zabrinskie to Silvano Arieti). In *Correspondence, 1940–1981.*
SAP (1948). L, June 17 (American Cancer Society to Silvano Arieti). In *Correspondence, 1940–1981.*
SAP (1953). L, January 17 (Silvano Arieti to M. Michaels). In *Correspondence, 1940–1981.*
SAP (1961). L, August 13 (Elio Arieti to Silvano Arieti). In *Correspondence, 1940–1981.*
SAP (1963). L, September 14 (Vincent Howles to Silvano Arieti). In *Correspondence, 1940–1981.*
SAP (1965). L, October 17 (Roman Rossleigh to Silvano Arieti). In *Correspondence, 1940–1981.*
SAP (1969a). L, April 11 (John Warkentin to Silvano Arieti). In *Correspondence, 1940–1981.*
SAP (1969b). L, April 24 (Silvano Arieti to John Warkentin). In *Correspondence, 1940–1981.*

SAP (1969c). L, April 29 (John Warkentin to Silvano Arieti). In *Correspondence, 1940–1981.*
SAP (1970). L, August 14 (Vin Rosenthal to Silvano Arieti). In *Correspondence, 1940–1981.*
SAP (1971). L, April 12 (Silvano Arieti to Christian Müller). In *Correspondence, 1940–1981.*
SAP (1972). T (Silvano Arieti, *My Chance and Luck in Crossing Dr. Bigelow's Path*).
SAP (1973a). T (Silvano Arieti, *The Meaning of Madness*). In *Speeches and Writings, 1940–1981. Lectures.*
SAP (1973b). T (Abramo Tuscany, *Swiss Oasis*). In *Speeches and Writings, 1940–1981. Plays.*
SAP (1974). L, May 21 (Silvano Arieti to Marie Coleman Nelson). In *Correspondence, 1940–1981.*
SAP (1975). L, 1975, October 28 (Silvano Arieti to Steven Moll). In *Correspondence, 1940–1981.*
SAP (1976a). L, May 5. (Paul Miller to Silvano Arieti). In *Correspondence, 1940–1981.*
SAP (1976b). L, May 25 (Silvano Arieti to P. Miller). In *Correspondence, 1940–1981.*
SAP (1976c). T (Paul Miller, *An Interpretation of Arieti*). In *Correspondence, 1940–1981.*
SAP (1979a). L, October 8 (Stuart Blaustein to Silvano Arieti). In *Correspondence, 1940–1981.*
SAP (1979b). L, October 22 (Silvano Arieti to Stuart Blaustein). In *Correspondence, 1940–1981.*
SAP (1979c). L, November 10 (Paolo Decina to Silvano Arieti).
SAP (Ua). B (Requirements for Admission of Graduates of Foreign Medical Schools to the Examination of the State Board of Medical Examiners).
SAP (Ub). HN, Unt. (Elio Arieti's notes). In *Subject File, 1914–1981. Personal. Arieti, Elio, 1914–1969.*
SAP (Uc). HN, Unt. (Poem Dedicated to Elio Arieti). In *Subject File, 1914–1981. Education. Workbooks, circa 1930.*
SAP (Ud). HN, Unt. (Notes on schizophrenia). In *Speeches and Writings, 1940–1981. Miscellany. Notes.*
SAP (Ue). HN, Unt. (Notes on schizophrenia). In *Speeches and Writings, 1940–1981. Miscellany. Draft. Undated.*
SAP (Uf). L (Silvano Arieti to Giacomo). In *Subject File, 1914–1981. Education. Workbooks, circa 1930.*
SAP (Ug). L (Silvano Arieti to L.E.). In *Correspondence, 1940–1981.*
SAP (Uh). L (Elio and Ines Arieti to Silvano Arieti). In *Correspondence, 1940–1981.*
SAP (Ui). M (Silvano Arieti, *Il Rabbino di Kleinay*). In *Subject File, 1914–1981. Education. Workbooks, circa 1930.*
SAP (Uj). M. Unt. (poem dedicated to Giacomo Dirindelli). In *Subject File, 1914–1981. Education. Workbooks, circa 1930.*
SAP (Uk). M (Silvano Arieti, *Illusione*). In *Subject File, 1914–1981. Education. Workbooks, circa 1930.*
SAP (Ul). M (Silvano Arieti, *A Vittorio Righi*). In *Subject File, 1914–1981. Education. Workbooks, circa 1930.*
SAP (Um). T (*Curriculum Vitae of Sylvan Arieti*). In *Subject File, 1914–1981. Employment. Curriculum Vitae, 1940–1980 circa.*
SAP (Un). T (Gerard Chrzanowski, *Profile: Silvano Arieti*). In *Subject File, 1914–1981. Personal. Profile, undated.*
Staum, M. (2016). *Cabanis. Enlightenment and Medical Philosophy in the French Revolution.* Princeton University Press.
Thompson Arieti, M. (1969). Silvano Arieti. A Profile. *Newsletter. Society of Medical Psychoanalysts,* 11 (3), 43–44.
Triarhou, L. C. (2021). Giovanni Mingazzini. *Journal of Neurology,* 268 (4), 1558–1559.
Werner, H. (1940). *Comparative Psychology of Mental Development.* Harper & Brothers.
Yerkes, R. M. (1932). Autobiography of Robert Mearn Yerkes. In C. Murchison (Ed.), *History of Psychology in Autobiography. Vol. II* (pp. 381–407). Clark University Press.
Yerkes, R. M., Morgulis, S. (1909). The Method of Pavlov in Animal Psychology. *The Psychological Bullettin,* 6 (8), 257–273.

Chapter 3
New World

3.1 Explorations

After having tried unsuccessfully to help Arieti leave Pilgrim State Hospital, in 1945 Newton Bigelow advised him to be patient, for, despite the "freezing" because of the war, "the way things are going in Europe, it may be that that difficulty will be solved of itself" soon (SAP 1945a).

Bigelow was right. The war was finally coming to an end. A few days after Bigelow wrote the letter, American troops entered German territory. Meanwhile, the Japanese battlefront turned into a conflagration. In April, the release of prisoners from Buchenwald, Bergen Belsen, and Dachau began. The Soviet Army entered Vienna. In Italy, Mussolini was captured and executed; in Germany the generals negotiated surrender and Hitler committed suicide. Events reached their climax with the bombs dropped on Hiroshima and Nagasaki. At last, in September, the war ended.

With the end of World War II, 1945 brought a great change in the lives of everyone. For Arieti a new phase began, with the birth of his first child, David (SAPf; SAP 1945c),[1] and radical changes in his professional life. With the end of the war, U.S. government authorities put an end to the rule "freezing" employees where they were, and he could finally leave Pilgrim State Hospital. He settled in New York, where he began his practice as a consultant at various hospitals—the Vanderbilt Clinic, with which he had been communicating since 1943 (SAP 1945d); Beth Israel Hospital, where he visited neuropsychiatric patients on Wednesday evenings (SAP 1946a); the Veterans Administration office on 7th Street, where he performed part-time service for a yearly salary of 3.600 dollars (SAP 1946d, 1946e); the Columbus Hospital, a charitable institution mainly serving Italian immigrants (SAP 1949d, 1950b); the Home for Aged Hebrews on 105th Street, which engaged him thanks to Rabbi David de Sola Pool, the spiritual leader of the Spanish and Portuguese Synagogue, where

[1] David was born on June 20, 1945. His date of birth is reported in the divorce agreement between Silvano Arieti and Jane Jaffe. On July 5, 1945, Robert M. Yerkes wrote: "To the Arieti's, the cordial greetings and congratulations of Doctor and Mrs. Yerkes, who wish to welcome enthusiastically David F. Arieti".

R. Passione, *Psychiatry and the Human Condition*, Springer Biographies,
https://doi.org/10.1007/978-3-031-09304-3_3

Silvano and Jane were married (SAP 1945b, 1946c). At the same time, he continued his labors as an unpaid research assistant at Columbia University. By now, he had published thirteen papers (Arieti 1941, 1942, 1943, 1944a, 1944b, 1944c, 1944d, 1945a, 1945b, 1945c, 1946; Ferraro & Arieti 1945; Minnick, Warden & Arieti 1946).

In 1946 Arieti opened an office at 1107 Park Avenue, on the corner of 89th Street in Manhattan (SAP Ud). The move from the Bronx, where he had registered his license in 1942, was a big change (SAP 1946b).[2] As his first secretary pointedly stated, he seemed to have moved "to the other side":

> Frankly, I think that with all your understanding of the human mind, you are a member of a richer class, and ... as part of a more secure class ... you cannot help but minimize the importance of social context in personality disturbance. ... We live in such a sick world that psychiatry can make as much progress as Sisyphus. Of course you are a physician and not a politician, but it's my belief that politics is the real physic for our times, and your efforts can have only local interest and efficacy for a small section of professional and affluent people (SAP 1946h).[3]

This statement mirrors the relevant change that had occurred in Arieti's professional and personal life—from his precarious past as an immigrant into a stable and more favorable condition. This evolution was reflected also in his change of residence: in 1947 he left the Bronx and settled in Brooklyn, at 292 Clinton Street (SAP 1947h). As he wrote to a former colleague from Pilgrim State Hospital, things began to go well (SAP 1947b, 1947d).

Also, after the war Silvano Arieti could finally reunite with his parents. In 1947 Elio Arieti sailed to United States to see his son (SAP 1947f, 1947g), and the following year was his mother's turn, as Ines boarded the "Saturnia" to arrive shortly before the birth of her second grandson, James (SAP 1948b, 1948c). A former school friend with whom Silvano reconnected by letters, Aldo, wrote to him: "I can imagine your joy in seeing your good mama again, after so many years" (SAP 1948a, my translation). At about the same time, he reconnected with other former close Italian friends, learning with great sorrow that one of them had died in the war (SAP 1948e).

Professionally, Arieti's private practice was still limited to psychiatry and neurology, and did not yet include psychoanalysis, a field in which he still lacked proper training. First his Italian academic education and then his full-time commitment at Pilgrim State Hospital, had prevented his training in dynamic psychiatry, a knowledge of which he had obtained, until then, only by his own efforts and by personal readings.

Nevertheless, now that his situation had changed, he could finally work to fill this gap. Driven by the desire to explore the work of Freud, at first he turned to the New York Psychoanalytic Institute, where the teachings of the father of psychoanalysis were fully accepted. Provided with the necessary letters of recommendation, in

[2] When he moved his office, Arieti also registered his medical license on the Register of Physicians and Surgeons of the County of New York.

[3] Dated December 5, 1946, this letter is unsigned, but from its content we learn that it was written by Arieti's secretary as an "advance notice" of her resignation. We also learn that they had met at Beth Israel, where the woman was Arieti's patient.

1946 he applied to the Institute for training in psychoanalysis (SAP 1946g, 1946i).[4] Unfortunately, in March 1947 his application was rejected (SAP 1947a).[5]

Thus, not everything was going smoothly for Arieti in the new course of his life. Shortly afterwards, he received the notice of a second rejection, from New York Medical College, where he had applied for a teaching position at the Psychiatric Department of post-graduate studies (SAP 1947c). Again, Silvano Arieti did not lose heart; he applied elsewhere and obtained a teaching position at the Long Island College of Medicine, where Howard Potter, who once had worked at the New York Psychiatric Institute, was now the Head of the Department of Psychiatry (SAP 1946f, 1957a, 1959a, 1959b). Besides teaching, Arieti's position included also clinical work in the Out-Patient Clinic. At the College, however, Arieti did not receive any payment for his service (SAP 1947e, 1954d).

For psychoanalytic training Arieti turned to the William Alanson White Institute, the headquarters of the cultural and interpersonal school of dynamic psychiatry promoted by two of the founding fathers of the Institute, Erich Fromm and Harry Stack Sullivan (Stern 1995; Conci, Dazzi & Mantovani 1997; Conci 2012; Shapiro 2017; Thompson 2017).

In a noted series of lectures in 1939, Sullivan had put "the study of processes that involve or go on between people" at the center of psychiatry. "The field of psychiatry," he wrote, "is the field of interpersonal relations, under any and all circumstances in which these relations exist" (Sullivan 1955, p.10).[6]

Sullivan's conceptions finally provided a theoretical framework for Arieti's observations at Pilgrim State Hospital, where he had ascertained the role played by interpersonal relations in the so-called "spontaneous recovery" of some patients:

> In the time spent in Pilgrim State Hospital ... I made what seemed to me interesting observations. I discovered that a few patients who ... had been considered hopeless would apparently recover or improve enough to be discharged, at times after many years of hospitalization. At that time these were considered cases of "spontaneous recovery." I was not satisfied with this explanation and looked more deeply into the matter. I soon discovered that these so-called spontaneous recoveries were not spontaneous at all, but the result of a relationship which had been established between the patient and an attendant or a nurse. ... I was impressed by the fact that even an advanced schizophrenic process had proved to be reversible or capable of being favorably influenced by a human contact. ... I had learned that whatever benefit the patient could receive had to come from his bonds with at least another human being (SAP 1977).

At that time, Arieti did not understand the cause of these spontaneous recoveries, not even with help from Bleuler's writings. Nor had Sigmund Freud been of help,

---

[4] Arieti was recommended by Harry I. Weinstock and Renato Almansi, both members of the Institute.

[5] The letter of rejection read: "Dear Dr. Arieti, your application for admission … has been studied with great care. From among the large number of applications received we are obliged to select only a small number. Our selections are based on various considerations, including the extent of previous psychiatric training, psychologic attitude, maturity and other factors. We regret to say that after thorough deliberation we cannot accept you as a student".

[6] The first edition of Sullivan's book was published in 1940.

given his strong pessimism about the efficacy of psychoanalysis with schizophrenic patients (Molaro 2013). Instead, it was *the patients* who showed Arieti the way, a way that he could finally focus on now, after many years, thanks to the perspective of interpersonal psychiatry.

In 1947 Arieti entered the William Alanson White Institute for the first time, then headed by Clara Thompson (Green 1964; Moulton 1986). Here he met Janet Rioch (SAP 1975),[7] who was "impressed with Silvano's never ending curiosity" and decided to take him under her supervision (SAP Ue).

At the Institute Arieti was introduced to a world brand new for him, that of "psychoanalytic dissidence," an alternative to the orthodox Freudian views (Kwawer 2017). The White Institute, for example, did not embrace libido theory and the Oedipal theory. Moreover, the Institute rejected both the idea that psychoanalysis did not work with psychotics and the traditional view that analysts must have a neutral attitude towards their patients.

Frieda Fromm-Reichmann was among the Institute's more experienced and authoritative teachers. Like her ex-husband Erich Fromm, she came from Germany as a refugee from Nazism. She arrived in the United States in 1935 and immediately found a job at the Chestnut Lodge, directed by Dexter Bullard, a psychoanalytically oriented psychiatric hospital in Rockville, Maryland, where she settled until her death in 1957. She lived in the cottage built for her in the clinic's park, and she spent her days with psychotic patients, devoting her energies to the psychotherapy of schizophrenia (Fromm-Reichmann 1950, 1959a; Hornstein 2000).

Her teaching was of the utmost importance for Arieti, who admired her pluralist approach that combined the lessons of four different figures—Kurt Goldstein, with whom she had been graduated in Germany; Sigmund Freud ; Georg Groddek, whose therapeutic optimism she greatly appreciated; and Harry Stack Sullivan, with whom she had collaborated in founding the William Alanson White Institute (SAP 1968; Arieti 1968).

Beyond Fromm-Reichmann's theories and science, Arieti was won over by her humanity and her conception of illness as a dramatic feature of the human condition. It was in this part of her work that he felt he had finally found what he was seeking—the idea that "the numberless ways, the infinite nuances with which people love or hate, help or hurt one another, in no other condition can be better observed than in the study of the schizophrenic disorder" (SAP 1968).

Arieti was particularly moved also by Fromm-Reichmann's thinking on loneliness (Fromm-Reichmann 1959b) as a multifaceted human experience that schizophrenia could reveal in its purest, dramatic, and most magnified form:

> Fromm-Reichmann was among the first to emphasize that the schizophrenic is not only alone in his world, but also lonely. His loneliness has a long and sad history. Contrary to

---

[7] About Janeth Rioch, Arieti wrote: "Dr. Janet Rioch has played a prominent role in my psychoanalytic training. She was the first one to interview me when I applied for training at the William Alanson White Institute. She was my teacher and supervisor. I have always thought of her with devotion and a spirit of friendship because, in addition to being an excellent teacher, she was able to radiate warmth and encouragement".

what many psychiatrists used to believe, the patient is not happy with his withdrawal but is ready to resume interpersonal relations, provided that he finds a person who is capable of removing the suspiciousness and distrust that originated with the first interpersonal relations and made him follow a solitary path (Arieti 1979b, p. 492).

Magnifying and making visible a typical feature of the human condition, a schizophrenic's loneliness raised questions that concern everybody. Arieti asked, "What does the cry of loneliness mean to any human being?" (Arieti 1968, p.88). He dwelled at length on the question, also in his correspondence with Frieda Fromm-Reichmann. "Incidentally," he wrote to her, "it occurs to me that the word *loneliness*, which is so important in English, has no equivalent in French, Italian and Spanish languages. It would be interesting to speculate why" (SAP 1956g).

More than all his other training, Arieti's work at the William Alanson White Institute enabled him to understand "that one becomes a person by virtue of relations with other human beings and not of inborn instinctual drives" (Arieti 1968, p. 86). At the Institute, he experienced first-hand the role played by the "relatedness" that later he would place at the center of the origins and treatment of schizophrenia (Arieti 1960). The interpersonal approach thus acquired for Arieti both a scientific and a personal meaning. And—to linger a moment longer on the merging of science and life—it was here, at the William Alanson White Institute, that he could finally overcome the multifaced feeling of loneliness that had no equivalent word in his native language but that he experienced personally as an immigrant in a foreign world.

3.2 Thinking in a New Language

At the William Alanson White Institute, Arieti befriended Gerard Chrzanowski, a psychiatrist of Polish origin who was portrayed as "Dr. Kik" in *The Snake Pit,* a famous book by Mary Jane Ward, his former patient (Ward 1946). Trained under Bleuler's supervision at Burghölzli, the Swiss psychiatric hospital that played a crucial role in the merging of psychoanalysis and psychiatry (Civita 2018), Chrzanowski arrived in the United States in late 1941 (Mann 2001). Arieti and Chrzanowski had in common the work experience at Pilgrim State Hospital, where both began to understand the importance of interpersonal relations in mental illnesses and their treatment.

Arieti's own interpersonal relations at Pilgrim State had not been easy, primarily because of his linguistic handicap, for he had arrived there with a still poor acquaintance with English. Thus, it is fair to say that while he was at Pilgrim his encounter with "the other" was two-sided: the United States with its English language on one side, and schizophrenia with its "word-salad" on the other. In both cases, Arieti understood that these languages were the expression of different ways of thinking, which he had to learn from the "outside," in the "New World" of the United States, and within the walls of the psychiatric hospital. It was by means of his difficult linguistic training among schizophrenics at Pilgrim State Hospital that he finally achieved a real mastery of the American language (Passione 2018).

We have here another example of the link between his life and science, since Arieti's interest in language and thought stemmed not only from his own experience, but also from the cultural background which preceded his education in psychiatry. In particular, the study of the eighteenth century philosopher Giambattita Vico's *Scienza nuova,* with its discussion of the link between the development of thought and the development of language, played a crucial role in Arieti's thinking (SAP 1977).

At the William Alanson White Institute, Arieti resumed and deepened these ideas with a plan of research initiated by Jacob Kasanin (Windholz 1947)[8] at a conference on language and thought in schizophrenia in 1939 in Chicago (Kasanin 1964a). The conference, attended also by Kurt Goldstein and Harry Stack Sullivan (Goldstein 1964; Sullivan 1964), had been a high-profile event, where pioneering ideas were discussed, including those of Lev Sëmenovič Vygotskij on conceptual thought in schizophrenia (Vygotskij 1934). Above all, it was a groundbreaking event, because in this area of research, ever since Kraepelin focused on cognitive aspects of schizophrenia with his concept of *dementia* (Berrios & Porter 1995), only a few scientific contributions had appeared. Though the work of Bleuler and Storch partially remedied this deficiency (Bleuler 1924; Storch 1924), the disturbances of associative processes in schizophrenia, as well as its typical "word-salad," continued to be thought to result from causes unknown but generally assumed to be of an organic and neurological kind. Moreover, as Kasanin remarked at the meeting, formal investigations in this field had been suspended, after World War I, also because of the success of dynamic psychiatry, which, following the teachings of Meyer, Freud, Jung and others, had focused on the emotional contents of schizophrenic symptoms rather than on the analysis of cognitive processes (Kasanin 1964b).

At the meeting, Harry Stack Sullivan took a step towards overcoming this lack of progress with a paper that dealt both with schizophrenic language and anxiety. In particular, he suggested that the schizophrenic adopted a highly individualistic, childlike, autistic language because of a need to counteract feelings of insecurity and to protect himself from the anxiety aroused by his interpersonal relations. Despite remaining on a mainly dynamic ground in his analysis, Sullivan was also suggesting something else—the connection of anxiety and language, which he described as a cognitive function related to communication and personal identity (Sullivan 1964).

Encouraged in person by Jacob Kasanin,[9] in 1947 Silvano Arieti took up the task set in motion at the 1939 meeting and followed Sullivan's suggestion with a cognitive analysis of anxiety aimed at identifying the underlying thought processes.

---

[8] Born in Russia in 1897, Kasanin arrived in the U.S. in 1915. In 1941 he took over the direction of the American Orthopsychiatric Association, an interdisciplinary and multi-professional organization committed to the promotion of mental health.

[9] Among Silvano Arieti's papers there is the excerpt of Kasanin's article entitled "Developmental Roots of Schizophrenia," published in 1945 in *The American Journal of Psychiatry*, with an encouraging inscription: "I enjoyed your paper 'Perceptual Alterations [in the Terminal Stage of Schizophrenia]' very much. Goes very much along with some [ideas of] mine".

His article, "The Processes of Expectation and Anticipation," in *The Journal of Nervous and Mental Disease,* was the outcome (Arieti 1947).

The article began with a critical analysis of the behavioral conception of anxiety as a result of conditioning. More specifically, Arieti referred to Pavlov's experimental research on neuroses that were aroused in animals by the application of a strong conditioned stimulus in cases where the animals were accustomed to a weak one (Hilgard & Marquis, 1940; Pavlov 1941). According to Arieti, this objective description, based on an extrapolation to humans of experimental data obtained in animals, did not acknowledge the irreducible human features of anxiety, which he, along with Sullivan, considered a powerful inner force of our species. In particular, Arieti suggested a preliminary distinction of two phenomena: *expectation,* which is common to humans and animals and can be described as "the capacity of the subject to anticipate certain events while a certain external stimulus is present," and *anticipation,* the typically human capacity "to foresee or predict future events even where there are no external stimuli which are directly or indirectly related to those events" (Arieti 1947, p. 471). Whereas experimental research on animals focused only on expectation, human anxiety, Arieti explained, was a "long-circuited" mechanism that entails anticipation—the mental representation of an event that has not yet taken place.

Thus, Arieti focused on inner processes that cannot be grasped in the study of animal behavior. Agreeing with Sullivan, he pointed out the distinctive nature of human anxiety; at the same time, he also took a step forward, as he tried to translate this dynamic perspective into cognitive terms; specifically, he analyzed the formal mechanisms of thinking that are at the basis of different psychopathological conditions connected to anxiety—the "psychotemporal inversion" in psychoneuroses, where it is "too painful for the patient to think of the uncertain future … and he prefers to withdraw into the past" (Arieti 1947, p. 478), and the psychotemporal restriction to the present time in advanced stages of schizophrenia, which often shows a pronounced loss of anticipation and a severe regression to the level of mere expectation.

Besides the formal and structural analysis of thought processes related to anxiety, Silvano Arieti utilized a developmental frame of reference that enabled him to build a bridge from neurology to child psychology and psychoanalysis:

> Anticipation, and therefore long-circuited anxiety, appear in man during the second year of life, or anal stage. However, during the first year, or oral stage, expectation and short-circuited anxiety are present. The baby expects the parents, or their substitutes, to relieve him from the tension produced by hunger, thirst, loneliness or other uncomfortable sensations.
>
> Anticipation originates during the anal period. At this stage of development, the child becomes able to postpone immediate pleasure for some future gratification. In other words, it is at this stage that the 'reality principle' originates. The child learns to give up his tendency to relieve the tension of his bowels, and postpone the act of defecation, in order to obtain the approval or love of his parents ... In the adult stage this future anticipation occupies the greatest part of man's thoughts, and consequently determines the greatest number of man's actions. It is to this process of anticipation that such phenomena or institutions as religion, life insurance, armament, etc., owe their origin and development (Arieti 1947, pp. 474–75).

Therefore, through observations on ontogenetic development, Arieti moved toward considering phylogenesis, where anticipation emerged when the stage of development characterized by present-oriented activities, such as hunting, was left behind for other future-oriented activities—two stages of human development that had already been described in the anthropological field by Géza Roheim and had been discussed by the philosopher Bertrand Russell (Roheim 1934; Russell 1945):

> Bertrand Russell describes well this differential characteristic between what he calls the savage man who is interested only in present problems, and what he calls the civilized man who is able to foresee the future: the civilized man is distinguished from the savage mainly by prudence, or, to use a slightly wider term, forethought. He is willing to endure present pains for the sake of future pleasures ... This habit began to be important with the rise of agriculture; no animal and no savage would work in the spring in order to have food next winter ... True forethought only arises when a man does something which no impulse urges him, because his reason tells him that he will profit by it at some future date. ... Russell thus describes the shifting from what is psychoanalytically known as the pleasure principle, to the reality principle (Arieti 1947, pp. 475–76).

Arieti's article shows the same integrative approach that characterizes his previous work, as he brought together different observations taken from different areas—from his clinical experience and from psychoanalytic theory; from anthropology and philosophy; from animal and comparative psychology (Yerkes & Yerkes 1928); from neurology, as he referred also to Malmo's studies on frontal lobes (Malmo 1942); and from the data available in psychosurgery (Freeman & Watts 1942), which showed that lobotomy caused a severe impairment of the ability "to foresee the results of a series of planned act" (Arieti 1947, p. 477) and therefore to project oneself into the future. In this respect, again, Arieti's psychiatry proves to be *polyphonic*, aimed at an interpretative synthesis of data and observations from different disciplines.

Those pages were not the result of a merely conceptual endeavor, since Arieti did not neglect clinical work for the sake of speculation and theory. After his experience at Pilgrim State Hospital, he maintained his skills as a clinician with his work in private practice and hospital consulting, as well as with his service at the Outpatient Clinic of the Kings County Hospital Center, where he began to work in 1948 (SAP 1948h).[10]

Schizophrenia continued to be his principal field of research, with a particular focus on schizophrenic thought, the peculiar structure of which he analyzed in his next article, "Special Logic of Schizophrenic and Other Types of Autistic Thought," published in 1948 in *Psychiatry*, a journal connected to the William Alanson White Institute (Arieti 1948). The article's title is significant, for, according to Arieti, schizophrenic thinking has its own *special logic* that differs fundamentally from classical logic. It follows unique rules and has a unique structure, the study of which requires us to reject the assumption that it is senseless and instead to learn its different language, its different frame of reference:

> The schizophrenic does not think with ordinary logic. His thought is not illogical or senseless, but follows a different system of logic which leads to deductions different from those usually

[10] The Kings County Hospital Center was connected to the Long Island College of Medicine, where Arieti held a teaching position.

> reached by the healthy person. The schizophrenic is seen in a position similar to that of a man who would solve mathematical problems not with our decimal system but with another hypothetic system and would consequently reach different solutions. In other words, the schizophrenic seems to have a faculty of conception which is constituted differently from that of the normal man (Arieti 1948, p. 326).

To understand a schizophrenic it is necessary to think the way he does. According to Arieti, this is the only method we have to grasp a comprehensible meaning from an apparently senseless word-salad. For this purpose, the study of emotional and unconscious content in delusional ideas is not sufficient. As suggested by Freud and Jung, a dynamic approach may allow us to understand the emotional forces lying behind delusions (Freud 1938; Jung 1936), but it does not help us to clarify the peculiar structure of schizophrenic thought nor to explain the change that "has occurred in the logic power of the patient so that he is not able any longer to test reality" (Arieti 1948, p. 326).

New interpretative tools were therefore needed to build on this foundation, and Arieti found them in the work of Eilhard von Domarus, a German psychiatrist who had arrived in the United States soon after receiving his medical degree, never to return to Europe. In his career, von Domarus dealt mainly with the relationships among logic, anthropology, and psychiatry (Arieti 1967). He was one of the participants at the meeting held in Chicago in 1939, where he had used the expression "paralogical thinking" in reference to a type of reasoning very different from the Aristotelian classical logic, a type that rather resembled primitive thought (von Domarus 1964). Taking a cue from this conception about schizophrenic reasoning, Arieti combined it with the idea of schizophrenia as a regression to primitive levels of development, and renamed von Domarus' *paralogic* as *paleologic:*

> Paleologic is to a great extent based on a principle enunciated by Von Domarus ... which, in slightly modified form, is as follows: *Whereas the normal person accepts identity only upon the basis of identical subjects, the paleologician accepts identity based upon identical predicates.* For instance, ... suppose that the following information is given to a schizophrenic: "The President of the United States is a person who was born in the United States. John Doe is a person who was born in the United States." In certain circumstances, the schizophrenic may conclude: "John Doe is the President of the United States." This conclusion, which to a normal person appears as delusional, is reached because the identity of the predicate of the two premises, "a person who was born in the United States," makes the schizophrenic accept the identity of the two subjects, the President of the United States and John Doe (Arieti 1948, pp. 326–27).

According to this principle of paleologic, then, the schizophrenic could be described as someone who confuses similarity with identity; his thinking is characterized by the loss of the essential distinction between similarity and difference.[11] In short, schizophrenic logic is governed by a mechanism that identifies subject

[11] Besides this basic principle, Arieti's article provided a detailed analysis of many other features of the schizophrenic logic: (1) schizophrenic logic is characterized by a reduction of the connotative power of a verbal symbol, so that it does not represent a group or a class or a concept, but only the specific object under discussion; (2) schizophrenic logic is characterized by an emphasis on verbalization and the schizophrenic focuses on verbal expressions, i.e. on words as words, not as symbols; (3) schizophrenic logic is characterized by an animistic and anthropomorphic concept of

and predicate, and this identification explains the bizarre language of these patients: "since the same subject may have numerous predicates, it is the choice of the predicate in the paleologic premises which will determine the great subjectivity, bizarreness, and often unpredictability of autistic thinking" (Arieti 1948, p. 328).[12]

> For instance, a patient of von Domarus' thought that Jesus, cigar boxes, and sex were identical. Study of this delusion disclosed that the common predicate, which led to the identification, was the state of being encircled. According to the patient, the head of Jesus, as of a saint, is encircled by a halo, the package of cigars by the tax band, and the woman by the sex glance of the man (Arieti 1948, p. 327).

In addition to von Domarus' work, Arieti took inspiration from other anthropological theories about primitive thinking, such as Lévy-Bruhl's concept of pre-logical thought, which Arieti considered a precursor of paleologic (Lévy-Bruhl 1910; Lévy Bruhl 1922). Useful hints came also from Max Levin's comparative analysis of the thought of schizophrenics and young children, as both were characterized by an inability to distinguish adequately between a symbol and the object symbolized (Levin 1938).

Influenced by reading Langer and Cassirer (Langer 1942; Cassirer 1946, 1953), in other articles Arieti reaffirmed this relevance of the symbolic dimension and its impairment in schizophrenia, and described the schizophrenic as someone who progressively loses the ability to think in terms of symbols and concepts (Arieti 1950a), for he is lodged on a level of "concrete thinking" similar to that described by Kurt Goldstein in patients with brain injury (Goldstein & Scheerer 1941).

Arieti's analysis of schizophrenic thinking was innovative from at least two points of view: first, Arieti brought disparate areas of research into a comparative and evolutionary approach, one that also owed a debt to the work of Heinz Werner (Werner 1940; Wapner & Kaplan 1964; Pea 1982); second, he combined a structural analysis of schizophrenic thought with the dynamic approach, since von Domarus' principle permitted Arieti to explain not only schizophrenic thinking, but also those oneiric distortions described by Freud in *The Interpretations of Dreams*. "Von Domarus' principle does not explain only schizophrenic thought, but also anything related to the primary process … I realized that von Domarus' principle was the law that Freud had been looking for, without succeeding in discovering it. I connected their contributions and realized that also in schizophrenia the laws of primary process were mechanisms set in motion by dynamic factors" (Arieti 1967, p. 245, my translation].

causality; (4) schizophrenic logic is characterized by the restriction of the psychotemporal field to the present time.

[12] In the same paper Arieti pointed out that schizophrenic thinking subverted the first three Aristotelian laws of thought—the law of identity, the law of contradiction, and the law of the excluded middle. On the other hand, schizophrenic thinking maintained the fourth law, that of sufficient reason, though in an altered form, since the schizophrenic refers causality not to external events but to psychological explanations in his inner world. External, physical causality is replaced by psychological and magical causality: "whereas the normal person is inclined to explain phenomena by logical deductions, often implying concepts involving the physical world, the autistic person, as well as the primitive and the child, is inclined to give a psychological explanations to all phenomena".

"Special Logic" was surely an important step towards that integration of dynamic and cognitive analysis that Jacob Kasanin and Harry Stack Sullivan had promoted at the Chicago meeting in 1939. It is perhaps for this reason, therefore, that Arieti decided to submit the article to *Psychiatry*, sending it to Sullivan in person (SAP 1948f). The journal enthusiastically accepted it. In 1948 Helen Tepper, Managing Editor of *Psychiatry*, wrote to Arieti: "I am happy to inform you that your paper has been accepted for publication in the current issue of *Psychiatry* … The editorial staff was enthusiastic about your contribution. One reader said: 'This paper is in all respects one of the finest which has come to me since I have been reading for *Psychiatry*.' I am certain that our readers will enjoy your important contribution" (SAP 1948g).

"Special Logic" was published with success. With its innovative character, it quickly became the emblem of a new course of psychiatric research. The article was applauded by von Domarus and Leo Kanner, a pioneer in the field of infantile autism, and was also soon adopted in courses at the Analytic Institute of Baltimore (SAP 1949a, 1949b, 1950d).

Thus, 1948 was a watershed year for Silvano Arieti, both from a professional and a personal point of view, for the scientific recognition he received and for the birth of his second son, James (SAP 1948d). Arieti's professional and social ascent is also evident in his change of residence, as he moved from Brooklyn to Manhattan, where from 1950 he lived on 545 West, 111th Street. He also moved his office, first to 10 East, 76th Street (SAP 1950e), and then to 22 East, 72nd street (Fig. 3.1).[13]

His private practice, study, and research, intensified. He worked very hard. Shortly afterwards he published, again in *Psychiatry*, a paper revisiting the classical Freudian conception of wit in light of von Domarus' principle (Arieti 1950b). Here Arieti describes in detail how the confusion between similarity and identity in schizophrenic thought is also a key to understanding humor, which, according to him, arises when a hearer of a joke realizes that he has wrongly identified two different, albeit similar, things. "For instance, a child who uses big words like a grown-up makes us smile because at first we react to him as to a man and then we realize that he is not a man. In a certain way, however, we feel that there is a man in him; this feeling is furthermore reinforced because we know that a child is potentially a grown-up. We're deceived until we focus our attention on the fact that he is a child" (Arieti 1950b, pp. 59–60).

Based on clinical observation of his patients' ostensible witticisms—"ostensible" since actual schizophrenics showed no sense of humor at all—this paper is a masterful combination of scientific analysis, philosophical reflection, and literary references, as Arieti places side by side the ideas of Freud, Jackson (Jackson 1887), von Domarus, Plato, Aristotle, Kant, Leibniz, and Herbert Spencer, among others. He also refers to James' philosophical thought on identity and similarity (James 1911), and to William Shakespeare's *Comedy of Errors*.

[13] Silvano Arieti's correspondence indicates that he moved his office from 76th street to 72nd street in 1958.

Fig. 3.1 Arieti's office in Manhattan, New York City (SAP, Library of Congress, Manuscript Division) © Courtesy of James Arieti—All rights reserved

In 1950, upon the acceptance of this article for publication in *Psychiatry*, the journal's managing editor wrote to Arieti:

> I have sent your manuscript to the printer's today. We had to do quite a job of Englishing on it ... You will notice that we made *paleologic* a noun and used paleological as an adjective. This corresponds to logic and logical. Since your use of the word is coined, more or less, we thought it wise to avoid entirely any confusion with the existing word in the dictionary which does not have the same meaning (SAP 1950a).

Thus, with this article Arieti finally became the acknowledged coiner of a new word. A new psychiatric language was being born.

During his years as a student at the William Alanson White Institute, Silvano Arieti also underwent personal analysis, at first with Eugene Eisner and later with Rose Spiegel (SAP 1949c). Finally, in 1952 he was awarded the diploma in psychoanalysis at the Institute (SAPd),[14] where he also started to take on an increasingly active role (SAP 1952b, 1952f, 1954b).

Tu sum up: by the early 1950s the previously disoriented, thin, almost ethereal young boy had reached a new solidity, one that also revealed itself physically in a heavier shape, as we can see in Figs. 3.2 and 3.3 (SAPg; SAP 1950c; Figs. 3.2 and

[14] In the Archive Arieti's diploma is erroneously dated 1953, the year of Arieti's graduation ceremony.

Fig. 3.2 Silvano Arieti in the mid-1940s (SAP, Library of Congress, Manuscript Division) © Courtesy of James Arieti—All rights reserved

Fig. 3.3 Silvano Arieti in the early 1950s (SAP, Library of Congress, Manuscript Division) © Courtesy of James Arieti—All rights reserved

3.3). With both a professional and personal solidity, Silvano Arieti became also a trusted and treasured support for his relatives and friends, in the United States and in Italy, with whom he had a close correspondence about personal and scientific issues. These were virtually uninterrupted dialogues in which he assumed the role of a most trusted "American psychiatrist" with whom all subjects could be talked about (SAPa). "When these troubles are past, know that you will feel younger than you do today" (SAP 1940, my translation), wrote his father in 1940. After ten years that time finally came.

3.3 American Psychiatrist

In 1950, the Long Island College of Medicine and the Kings County Hospital became part of a hospital network connected with the State University of New York, a place where many future psychiatrists would be trained. Arieti wrote to Howard Potter,

Head of the Department of Psychiatry, asking to teach a new psychoanalytic seminar on the *interpretation of schizophrenia*: "in each session," he wrote, "after a short theoretical introduction, there would be the presentation of a case … The discussion would be about the psychopathology and dynamics of schizophrenia; but the general point of view would be eclectic" (SAP 1951). Aimed at the psychoanalytic education of physicians, the initiative was inspired by what Sándor Rado had done at the medical school of Columbia University, where in 1944 he had launched the first academic school of psychoanalysis (SAP 1952a; Roazen & Swerdloff 1995; Tomlinson 2010).[15]

Arieti's request was approved, and in the fall of 1951 he inaugurated his seminar on schizophrenia for first-year residents (SAP 1952c; SAP Uf).

Compared with the Freudian approach that Arieti had once used to explain some regressive schizophrenic symptoms (Arieti 1944b, 1945a, 1945b), his theoretical frame of reference was now considerably different, as in his seminar he followed Sullivan's teaching and criticized the instinctualism of the father of psychoanalysis, whose major mistake was neglecting the role of interpersonal relations and anxiety in schizophrenia (SAP Uf). When in 1952 Edward Podolsky asked Arieti's permission to include his article "The Placing-Into-Mouth" in a book to be published by the Philosophical Library, Arieti agreed with the proviso that he be allowed to amend the text where he referred to Freud's libido theory: "Inasmuch I cannot fully subscribe any longer to Freud's libido theory, I would like to replace the sentences which are crossed with the following paragraph: *One does not necessarily need to subscribe to Freud's libido theory to acknowledge some kind of behavior which may be called oral. The author of this article prefers Sullivan's formulation of the oral complex, according to which the first vital experiences are chiefly connected with the not necessarily erotic impressions of the mouth-throat-larynx zone*" (SAP 1952d).

The seminar was a proper workshop for a "brief monograph on the subject of schizophrenia" that Arieti already had in mind to convey his original thinking on this illness. His book, *Interpretation of Schizophrenia,* was thus engendered by the seminar and conformed closely to the course syllabus (SAP 1951; SAP Uf).

In the summer of 1953 Arieti submitted his manuscript to Basic Books, but the publisher, not considering it a profitable project, rejected it (SAP 1953a). Arieti then submitted it to Robert Brunner, who accepted it (SAP 1953c, 1953f; SAP 1953g).

Arieti worked feverishly on the manuscript, often remaining in his office into the late night after treating patients during the day. His Aunt Yolanda worried about him from her home in Miami Beach, Florida, for she was concerned about his being by himself, "so isolated in this country, where crime seems to gain ground more and more, every day" (SAP 1953d, my translation). She suggested paying one of his students to spend the night studying in the waiting room to his office.

The manuscript was ready in January 1954, and Arieti sent it off (SAP 1954a). As Clara Thompson stressed in a letter to Frieda Fromm-Reichmann, the book, in its theoretical and therapeutical orientation, owed a great debt to the lessons taught

[15] Arieti and Rado were acquainted with one another. In 1952 Rado wrote to ask information about John Meth, who had applied to the Columbia's psychoanalytic school.

at the William Alanson White Institute (SAP 1954c). According to Arieti, however, the crucial role of interpersonal relations had already appeared as a basic principle in human sciences long before Sullivan, as shown in the nineteenth century debate in France about the wild child Victor de l'Aveyron (Gusdorf 1960; Moravia 1972, 1974). On this matter, when Arieti discussed his forthcoming book at a conference at the Sullivan Society of New York, he said:

> Even certain functions of man which seem due only to a native organic endowment, like speech and the power for any form of high symbolization, are not exclusively the result of a more evolved neurological apparatus. The new cerebral areas with which man is provided, like the language centers and the prefrontal lobes, are necessary for processes of high symbolization, but not sufficient. The other element which is equally important for the development of these functions is exposure to, or contact with, other human beings—children who were lost in the woods and managed to survive and who were found much later ... never learned to talk a language (Arieti 1954, p. 396).

Arieti placed his book in the tradition of Sullivan's modern psychiatry and George Mead's social psychology (Mead 1934), taking both interpersonal relations and socialization as key concepts for the development of self-identity and human nature at large. In schizophrenia, Arieti explained, we always find an early disturbance of self-image. A child, whose early self-image develops from how the adults who surround him see him, cannot bear the burden of the bad self-image conveyed by hostile and rejecting parents. To avoid the anxiety aroused by this primary relationship, the child tries to steer clear of socialization. He does not accept the symbols of his significant adults; instead, he withdraws into a private, separate world, as similarly described by Leo Kanner in his studies on infantile autism (Kanner 1944, 1946).A schizophrenic psychosis, in this way, would have a precise teleological meaning, since it aims at removing the devaluing internal self-image—the "bad me," Sullivan said (Arieti 1954, p. 406).

Arieti's conception of schizophrenia was thus indebted to the school of Sullivan and to dynamic psychiatry. *Interpretation of Schizophrenia*, however, went beyond a mere interpersonal and dynamic explanation to focus on the role played in schizophrenia by intrapsychic and cognitive processes. Arieti's idea was that there are no emotions without thoughts, no feelings without a concomitant intellectual content, and, for this reason, a further step beyond dynamic analysis was required:

> Let us take, for instance, the case of a deluded, moderately regressed schizophrenic, who think that she is a crocodile. If we succeed in probing the patient psychoanalytically we may determine what emotional forces are at play in this case and why the patient unconsciously wished to be a crocodile. ... Psychoanalysis will not explain, however, why the patient is absolutely sure that she is a crocodile in spite of the most complete contradictory evidence. ... It is obvious that a change has taken place in the logic functions of the schizophrenic, so that she is not able any longer to test reality. We mean a qualitative change, of course, and not a quantitative change in the sense that her I.Q. is lowered. To limit ourselves to say that the ego of the schizophrenic is disintegrating under the stress of the emotions is to cover the unknown with a terminological screen (Arieti 1950c, p. 4).

Emphasizing the role of cognition was a challenge in the America of the early 1950s, when behaviorism, instinctual Freudianism, and interpersonal perspective prevailed, and when examining the role of cognition was still looked upon with great suspicion (SAP 1979). Moreover, the cognition to which Arieti referred was not conceived in the same terms as "information theory" would be at the famous 1956 Symposium held at the Massachusetts Institute of Technology, which led to a cognitive revolution in a new science of the mind (Gardner 1985). Arieti's conception of cognition was not related to information theory, but referred to the close connection of feelings and thoughts that molded the structure of intrapsychic subjectivity.

As I have already explained, Arieti came to the subject of cognition by analyzing schizophrenic language, which he discussed at Clark University in 1953 at a Symposium on Expressive (Symbolic) Aspects of Language. The Symposium was organized by Heinz Werner, who invited Arieti as "one of the very few persons having a profound knowledge of psychopathological processes involved in linguistic thought" (SAP 1953b). The invitation, which arrived while Arieti was still working on his manuscript on schizophrenia, helped him to focus on the integration between a dynamic approach and a structural analysis of thought that he wished to convey in his forthcoming book. At Clark, where Arieti met Susanne Langer and befriended Bernard Kaplan, one of Werner's pupils, he elucidated how the schizophrenic word-salad was explanatory of this indispensable integration:

> As long as we try to interpret directly a sentence like "The house burnt the cow, horrendendously always," we do not get very far. However, as long as the patients use verbal symbols, although with paleologic distortions, we should not give up the hope of understanding them. ... The tendency to identify in the schizophrenic is so great that he identifies things and symbols which should only be associated. If we grasp this point, we understand some of the mysteries of word-salad. ... It remains to the perspicacity of the therapist and to his knowledge of the patient's life history, to find out what the elements stand for. ... Now let us go back to our patient who ... uttered the sentence "The house burnt the cow, horrendendously always." ... I had induced the patient to talk about the activities of the day. On a previous occasion she said something which had been interpreted by me as expressing resentment toward her mother, who now was doing all the housework, not to relieve the patient, but allegedly to imply that the patient was sick and unable to do it. I assumed thus that the word "house" stood for mother, mother being the vital part of the house. This identification is also common in dreams. "Burnt" stood for cooked, and the cow stood for meat or meal. "Horrendendously" is a neologism which, for reasons we do not know, the patient used to reenforce subjectively the social symbol "horrendously." The sentence probably refers to the activity of the mother who had prepared lunch before the patient came for the interview. The literal translation would be "the mother cooked the meal in a horrendous way, as she always does things." It is very doubtful, however, that the patient meant only this. In schizophrenic language we have always to contend with the syncretism or condensation which is also a characteristic of dreams and primitive languages. I am inclined to think that the patient meant also that the mother burnt, with anger, irritation, hostility, the patient, who is identifying herself with a submissive cow. At the same time that the patient wanted to convey a specific fact about preparing lunch, a total situational picture, the relation between her and her mother was superimposed. This sentence may be roughly compared to a photographic film which has been exposed twice, once when a particular snapshot was taken, and a second time when a landscape including the first scene was photographed. Schizophrenic language bristles with different planes of meaning (Arieti 1955a, pp. 62–64).

To decode the baffling schizophrenic language it was necessary to proceed on two different levels at the same time: on the one hand, to recognize the structural laws of schizophrenic logic; on the other, to interpret its elements. The integration between these two different levels—the double work of *translation* and *interpretation*—is a cornerstone of Arieti's book, the structure of which put in order a psychodynamic and then a structural analysis of the illness, connecting them with the concrete example of clinical cases (Arieti 1955b).

In short, the book describes schizophrenia as a regression to archaic cognitive levels that takes place under the pressure of the severe anxiety connected to the process of socialization, when an individual refuses common symbols and assumes an archaic and autistic way of thinking. Yet, this regression is progressive, tending to deepen more and more. Though it aims at avoiding anxiety, it actually ends up failing to reach its goal, thus bringing forth other disturbances that repeat the cycle, producing a further drift toward lower levels.

Interpretation of Schizophrenia was published in 1955. Shortly before, Robert Brunner's publishing house had been taken over by Arthur Rosenthal, the President of Basic Books, which thereby acquired a volume that the firm had rejected as a result of its dismal assessment of projected sales. Its great success proved Basic Books to have been mistaken; only three months after its publication its sales were so numerous that a second printing was scheduled (SAP 1955b, 1955g, 1955h, 1955l).

The book received great approbation. Von Domarus, who read the manuscript before its publication, considered it "a milestone" (SAP 1953e). Charles Wahl, from the University of California, Los Angeles, described it as "a monumental contribution to scientific thought" (SAP 1955i). Since Bleuler's work on schizophrenia, another reviewer noticed, nothing like this book had ever been seen (Liebert 1956). And so on (SAPc).

Silvano Arieti was invited to lecture throughout the United States (SAP 1955h). He became famous. The Columbia Broadcasting System invited him to serve as a high-profile consultant for a special program on mental illnesses (SAP 1955k). His name also became known abroad. Students, patients, and relatives of schizophrenics, impressed by his analysis of schizophrenia, wrote to him (SAPb).

With *Interpretation of Schizophrenia* Arieti became one of the most illustrious American psychiatrists, so much so that in 1956 his autographed photograph, his articles, and his book became part of a special collection of the Armed Forces Medical Library devoted to "contemporaries who have made significant contributions to the medical sciences" (SAP 1956f).

He repeatedly highlighted the American scientific environment in which he worked. In correspondence with Italian colleagues, for example, he insisted particularly on his debt to the New World (SAPa; SAPb; SAPh). His book, he wrote, could not have been written anywhere other than in the United States, where his thinking was fed not only by Sullivan's teaching—an "American psychoanalysis" (SAP 1955c, my translation), he pointed out—but also by the typical American liveliness of dynamic and psychological culture.

The American tradition behind the book was also recognized and welcomed in Japan, the first country to translate Arieti's work (SAPi). Ten years after the American bombing of Hiroshima and Nagasaki, the Japanese edition was the sign of an alliance in the fight against the common plague of schizophrenia, an illness that laid bare the most horrid picture of humanity estranged from itself and required for its healing a solid community of effort and good will—"the overcoming of all possible barriers which divide the human community" (SAP Ug).[16] In short, science was mightier than war.

In being an "American book," *Interpretation of Schizophrenia* had a frame of reference very different from that of Europe's phenomenological and existential schools, as revealed, for example, by the lack of bibliographical references to the works of Karl Jaspers, Ludwig Binswanger, and Eugène Minkowski. This feature of the book reflected the minor relevance of their approach for American psychiatry, which, however, largely owing to the influence of Rollo May, later started to open up to these schools (Villeneuve 1965; Spiegelberg 1972).

In this regard, one should consider the example of Arieti's reception of the work of Alfred Storch, a psychiatrist deeply influenced by Heidegger's philosophy and Jaspers' phenomenological approach. In 1922 Storch had published *Das Archaisch-Primitive Erleben und Denken der Schizophrenen,* the American edition of which translated the word *Erleben* ("lived experience") with *inner experiences* (Storch 1924). The switch from one culture to another, brought about by linguistic choices, caused the footprint of a well-defined theoretical and philosophical tradition to disappear. It is therefore not surprising that Arieti took Storch for a pupil of Heinz Werner, contextualizing his work in the field of comparative psychology rather than phenomenology (Arieti 1967).

Thus, Arieti was unfamiliar with the phenomenological and existential psychiatric schools, and, as revealed by his correspondence with Tullio Bazzi, an Italian colleague, had not yet approached Jaspers' *General Psychopathology* when he began work on *Interpretation of Schizophrenia* (SAP 1952e). Arieti, recognizing his lack of knowledge on this subject, later tried to fill the gap. In the 1960s, for example, he participated in the initiatives of the American Association of Existential Psychology and Psychiatry, where he met Rollo May and Medard Boss (SAP 1960b, 1960c). In 1960 he also tried to establish contact with Eugène Minkowsky, a leading figure of the phenomenological school (Minkowski 1933, 1970), who, as far as we know, never replied to the overture of his American colleague (SAP 1960d).[17] At any rate, as Paul-Claude Racamier pointed out, Arieti's approach was very different from Minkowsky's "purely phenomenological" point of view and from Binswanger's existential orientation, and, if anything, rather resembled the structural approach of Henry Ey (SAP 1959c).

An opportunity for a deeper exchange with existential psychiatric views came from Arieti's connection with Gaetano Benedetti, an Italian psychiatrist based in Basel, where he held the Chair of Mental Hygiene and Psychotherapy (Conci 2008; Dalle Luche, Giacobbi & Di Petta 2013; Di Petta 2014). His fame as a therapist

[16] These words were used in the English typescript of the Preface to the Japanese edition of the book.

[17] Arieti tried to contact Minkowsky through Joseph Gabel, a young French student who was close to the author of *Le temps vécu*. Gabel delivered Arieti's writings to Minkowsky, but Arieti never received any reply.

and pioneer in the treatment of schizophrenia induced Arieti to go to Switzerland, where in 1957 he attended the Second International Congress of Psychiatry in Zurich, meeting Benedetti in person. Before his departure he asked Daniel Blain, Medical Director of the American Psychiatric Association, to sponsor his participation at the congress, an institutional move that showed his intention to be officially recognized as an American psychiatrist before an international audience (SAP 1956a, 1956j).

From a scientific point of view, *Interpretation of Schizophrenia* was an important contribution to psychiatric knowledge; from a personal point of view it was the well-earned success of a former immigrant who had struggled for his professional rise. Armando Ferraro, who had supported him in his early hard days in America, welcomed Arieti's success with great joy. Upon reading the book, he wrote a touching letter to his former pupil. He said that as he pored over its pages he was strongly impressed with what he read and found confirmation of the exceptional skills he had detected in the young doctor during their years together at the New York Psychiatric Institute (SAP 1955d). Those years now seemed a remote past.

3.4 Borderland

With the publication of *Interpretation of Schizophrenia* Arieti was widely acknowledged as one of the leading experts in his field. From this point on, his readership identified him with his book, as shown in a 1975 review of "Arieti's second edition" (H.P.P. 1975, p. 10), where the reviewer used the pronoun *his* instead of *its* for the book's outline and contents. Nevertheless, the success was not without dissent. In its innovative character, the book was challenging and provocative, and could not help arousing some skepticism and a certain lack of agreement, sometimes veiled, sometimes openly hostile.

Its originality did not lie in the idea of the psychogenesis of schizophrenia, for in those years American psychiatry had already widely accepted dynamic conceptions (American Psychiatric Association 1969; Sabshin 1990, 2008). Nor was the book's innovation its emphasis on psychotherapy, since five years earlier there had appeared a milestone of this approach in Frieda Fromm-Reichmann's *Principles of Psychotherapy*. "When we come to practice and therapy, we must turn to your contribution," he wrote to her shortly after the publication of his book (SAP 1956d).

The real innovation of Arieti's endeavor consisted in his theoretical point of view, conveyed by means of a thorough survey of schizophrenia. The book offered an epistemological judgment aimed at an integration of a scientific analysis and a dynamic approach. "It is one of the main purposes of my research to integrate the organic and the psychodynamic approaches," Arieti wrote to Klüver shortly after the publication of his book (SAP 1955e).

The attempt to integrate these two perspectives was linked with the wider question of the scientific foundations of psychiatry, upon which Arieti dwelled during the months before the book's publication. Taking from Heinz Werner the concept of *syncretism* as a feature of the primitive mind, in a letter he thoroughly discussed with him its relation to psychiatry as science:

> Although syncretism applies particularly to primitive phenomena, in my opinion it applies to a certain degree to all phenomena in nature. Science may be viewed as a relentless attempt to undo this natural syncretism. In describing a psychological phenomenon I may adopt a non-scientific point of view and emphasize the syncretic character ... Incidentally, this is the point of view which we often follow in psychodynamic psychiatry. If, on the other hand, I want to follow the scientific method, I have to separate the elements ... I know that water is ... very different from the sum of atoms of oxygen and hydrogen. Gestalt psychology has taught me that. However, if I want to follow a scientific or nomothetic point of view, I have no choice: I must state that water is composed of oxygen and hydrogen.
>
> Similarly, the modern psychiatrist finds himself in a very difficult situation, inasmuch as he has to alternate rapidly or to see simultaneously the two points of view, if he wants to understand fully what is going on in the patient ... With the progress of science, things that we separate correspond less and less to natural things and become more and more parts, abstractions, symbols. Contrary is what generally assumed, not primitive thinking but scientific thought is the most symbolic. ...The scientific method almost changes the object into a concept and segments the whole phenomenon into symbols. But we have no choice: we must do that unless we give up the pragmatic goal of science (SAP 1955a).

Arieti seems here to suggest that the moment had arrived for dynamic psychiatry to overcome its typical primitive syncretism and to follow science's pragmatic goal with a more eclectic and integrative approach.

Soon afterwards, Jurgen Ruesch, a psychiatrist of Italian and Swiss origins (Balbuena Rivera 2018), also stressed the need for a new methodological and theoretical framework. He wrote:

> In the behavioral disciplines there exist two distinctly different factions ... The proponents of psychiatry as a science are positivists who want to follow the footsteps of physical science. Exponents of psychiatry as an art are subjectivists who maintain that ordinary scientific methods are entirely inapplicable to the data of behavior. They believe that the physical scientist's method of splitting the universe into progressively smaller entities is a meaningless procedure when applied to behavior. Together with the Gestalt psychologists, they emphasize that the comprehension of ... psychological phenomena and the understanding of human experience are based on the appreciation of patterns and that reduction, analysis, and isolation of elements destroy this patterning. Another significant difference pertains to the ... assumptions about the position ... of the human observer. Obviously a unified theory of human behavior has to be developed in which, in contrast to the schemes used in natural science, the observer is assigned a variable position within the system of which he is a part. Although the beginnings of such an undertaking are in the making, at present not enough effort, time, and money have been devoted to a revamping of the theoretical framework of psychiatry (Ruesch 1957, pp. 95–96).

"I agree fully with what you say and must congratulate you for exposing the situation so clearly and courageously," Arieti replied in a letter to Reusch (SAP 1957c). With its reference to an observer's position, Reusch's contribution straightforwardly called into question practical issues, for an epistemological consideration was not merely a theoretical matter but concerned crucial problems in the practice of psychotherapy, a field still considered by some "more an art than a science" (Arieti 1955b, p. 479). This matter, debated also by John Bowlby (Bowlby 1979), raised highly controversial and problematic questions. One question was when to terminate therapy with psychotic patients and how to establish criteria for doing so. Could objective goals be established ahead of time? According to Arieti, they could not, for in psychotherapy "not only is the patient unable to formulate goals, but even the therapist. … The real goals cannot be formulated early in the treatment of any patient" (SAP 1961). Prediction, though crucial in science, was therefore excluded, since in psychiatry one generally could not make predictions (Arieti 1957a). This troublesome evidence implied a need for a new thinking about the scientific status of psychiatry. To begin with, the adoption of a probabilistic epistemology would perhaps have been appropriate in the field, since the laws of human behavior seem not so indisputable as those of natural science.

After the publication of his book, Silvano Arieti went more deeply into questions that linked the problem of the integration of dynamic and scientific approaches to the relationship of history and nature in human beings (Arieti 1956a, 1957b, 1957c). Is an adult human the result of an original organic endowment that can be investigated solely by a quantitative approach, or is a human being a product of his experiences, to be examined qualitatively by a historical method? This question was also debated among anthropologists who emphasized culture and history and others who emphasized a biological approach focused on the concept of race. And in fact it was to anthropology that Arieti turned. "Kroeber," he wrote, "says that anthropology is both history and science. Psychiatry also should consist of the study of the history of the individual and of scientific examination of the phenomena to which the individual is subjected. The two methods *complement* and revitalize one another" (Arieti 1956a, p. 33). We find here the explicit expression of Arieti's complementary and integrative perspective (Passione 2016a, 2016b; Cerea 2015).

The difference between *Naturwissenschaft* and *Geisteswissenschaft* ("natural and human science") suggested by Dilthey, as well as the double methodology indicated by Windelband with the distinction between nomothetic and idiographic approaches, was to be taken seriously in the study of personality and its disorders. On the one hand, one should consider man as an *object* in the world of nature and try to *explain* the natural laws of his behavior; on the other hand, one should focus on the *subject* in his uniqueness and *understand* his individual history, which can never be reduced to the uniformity of general laws (Arieti 1957b).

In articles, Arieti stood his ground in his opposition to extremist schools of thought, rejecting both a purely biological and nomothetic approach (which brought forth few results in the field of biological psychiatry) and a purely historical and idiographic one, of which he found examples in the work of Ortega Y Gasset ("Ortega y Gasset goes to the extreme of saying that *that man has no nature; what he has is history*") and, to some extent, in the work of Erich Fromm ("He has paid little attention to the scientific-nomothetic … His emphasis on the uniqueness, on the individuality … are all features pertaining to the idiographic method.") (Arieti 1957b, pp. 536, 549; Ortega y Gasset 1936; Fromm 1951). According to Arieti, the scientific method had to reassert itself between these opposing idiographic and nomothetic conceptions: it could not be neglected by dynamic psychiatry nor be left at the mercy of "the pseudopsychology of behaviorism" (Arieti 1957b, p. 542).

Undoubtedly Arieti's thinking, peppered with philosophical references, stood out in the American psychiatric literature of those years: "I must candidly admit I was surprised that a psychiatrist had read Dilthey!" wrote Bruce Cameron of Bradley University (SAP 1960a).

This epistemological trait of Arieti's psychiatric thought emerged again in the article "The Two Aspects of Schizophrenia," in which he described schizophrenia as both a historical/experiential condition and a natural/extra-experiential one. Regarding the former, in addition to Harry Stack Sullivan's and Mead's views, he also valued those of Martin Buber, the "philosopher of otherness" (SAP 1979) whose work he had approached in his youth at Pardo Roques' gatherings in Pisa (Arieti 1979a), and whom he later met in person at the William Alanson White Institute (SAP 1956h).[18] As observed by a colleague in a letter dated 1954, Buber's philosophy seemed to proceed along the same path of Sullivan's and Mead's thinking about identity and human nature (SAP 1954e).

Buber, in his 1923 book entitled *Ich und Du* (translated into English as *I and Thou*), had defined the pair *I-Thou* as a "primary word": "there is no *I* taken in itself, but only the *I* of the primary word I-Thou" (Buber 1938, p. 4). Accordingly, no self-identity could ever develop without the *Thou*. "The person emerges from this relationship with the other, the *Thou*" (Arieti 1957c, p. 404).

Arieti considered schizophrenia as an experiential condition that was closely related to this view. The "schizophrenic cleavage" stemmed from experience insofar as it was a "never-complete acceptance of the *Thou*, or of the social self, or that part of the self which originates from others. This *Thou* which is not completely accepted or integrated, tends to … become dissociated, something like a foreign body … which is … externalized … in forms of projections and hallucinations" (Arieti 1957c, p. 405).

In addition to this experiential feature of schizophrenia, Arieti considered its extra-experiential feature, which pertained to natural phenomena and stemmed from the common biological endowment of man:

> In order to avoid anxiety, ... to reject the Thou-world (the Thou which is in the I and which causes so much distress), the schizophrenic ... withdraws from a way of reasoning which is logical and is shared by society, and adopts an archaic and paleological way of thinking. He also withdraws from a system of symbolization shared with his fellow human beings, and adopts his own private symbols ... The withdrawal is possible because the archaic or autistic mechanisms are within him as inborn potentialities, not as experiential phenomena. These mechanisms are part of his biological entity and are, therefore, susceptible to scientific studies (Arieti 1957c, p. 413).

Thus, the schizophrenic's refusal of the *Thou* that is part of the *I* entailed a paradoxical refusal of identity in the name of the self; or, perhaps more accurately, it entailed a refusal of the historical and social self in the name of the unhistorical and biological self. In this way, schizophrenic cleavage shed light on different facets of the original duality of human identity. On the one hand, it showed that humans belong to the world of nature and to the world of history; on the other, it showed the

---

[18] In 1957 Buber was invited to deliver the fourth William Alanson White Memorial Lecture. It was his first visit in the United States.

basic contradictions in humans, as the schizophrenic was someone who protected the consistency of his identity by a cleavage of his identity, a process that Buber described as a "self-contradiction" in which a person could experience the horror of his inner doppelganger. "Here," Buber wrote, "is the verge of life: flight of an unfulfilled life to the senseless semblance of fulfilment" (Buber 1938, p. 70).

The person who asserts himself by means of a denial of himself was a paradox that represented for Arieti one of the topics of major philosophical interest in the study of schizophrenia (Passione 2019), which, in its typical features revealed, as through a magnifying glass, a human condition burdened with dichotomies and contradictions—"a glimpse of inner struggles" (SAP Ua), as Arieti wrote in his notes. Thus, the basic dichotomy revealed in the study of the two aspects of schizophrenia concerned not only the schizophrenic, but human nature as a whole. It was an ancient dichotomy; "not the dichotomy between body and soul, which is obsolete today, but … a dichotomy between essence and existence and, one may add, between the universal and the particular, the form and the content … In other words, *it is the same dichotomy which is inherent in whatever pertains to the life of man*" (Arieti 1957c, p. 415).

Arieti's article, the contents of which were released before its publication, in lectures given in 1955 (SAP 1955f, 1955j), was received with a certain disfavor by the members of the William Alanson White Institute who adhered to the cultural perspective of dynamic psychiatry. Perhaps not surprisingly, *Psychiatry* rejected the article, claiming that it was "unsuitable" (SAP 1955m) for the journal.

Shortly after this rejection, *Interpretation of Schizophrenia* received a scathing review in the same journal. In it, Leslie Schaffer harshly criticized two major aspects of Arieti's work: the concept of *regression*, which Arieti took from neurological theories; and Arieti's acceptance of Aristotelian logic as the canon of normal thought—a position Schaffer considered authoritarian and normative (Schaffer 1956). Schaffer depicted Arieti's ideas in a caricatured, even hostile manner—so hostile that Bernard Kaplan, from Clark University, decided to write to the journal and to the William Alanson White Foundation to express his strong protestation against Schaffer's one-sided, highly biased critique of Arieti's book (SAP 1956i, 1957b; SAP Ub).

It is curious that the only truly bad review of the book came from the school to which Arieti himself belonged, but perhaps it revealed the unwillingness to accept Arieti's anti-relativistic approach and his emphasis on biological aspects of behavior by the members of a school that adopted a cultural approach to dynamic psychiatry.

At that time the problem of the relationship with the medical profession was an highly pressing issue at the Institute. It was going through infighting between medical and non-medical factions—i.e., between those who wished to limit membership at the Institute to medical graduates only, and those who wished to grant admission also to graduates in psychology. In March 1956, a proposal was made to discontinue the training of psychologists (SAP 1956e, 1956k, 1956l). Arieti was in favor of the proposal, explaining his rationale thus:

> I am fundamentally in agreement with the proposal ... I am teaching to the second year residents of the Kings County Hospital ... In this capacity I have become aware that many residents, even the best, are reluctant to apply ... at the W.A.W. Institute. One of the reasons

> given is that the Institute trains non-medical candidates. ... I assume that the same situation prevails in many medical centers ... I feel our efforts should be direct toward establishing stronger ties with medical schools and with the psychosomatic aspect of medicine in general. This aim cannot be achieved as long as we sponsor the cause of the psychologists (SAP Uc).[19]

So, despite his integrative and pluralistic approach, Arieti chose at this moment to emphasize boundaries instead of integration. His reasoning was perhaps inconsistent with his general views, but is explainable by the clearly stated strategic character of his position—removing all obstacles to the education of a new generation of psychiatrists. This was his primary goal at the Institute, where he was increasingly active (SAPe).[20]

The point is that Arieti understood well how difficult it was to be heard by medical specialists. He had experienced this deafness in the rejection of his article on the role played by the nervous system in schizophrenia, rejected both by *Neurology*, the journal of the American Academy of Neurology, and by *The Archives of Neurology and Psychiatry* edited by Roy Grinker, who worked at the Institute for Psychosomatic and Psychiatric Research of the Michael Reese Hospital, Chicago (SAP 1956b, 1956c). In fact, despite extending his hand to neurologists, in his article Arieti vigorously criticized their tendency to adhere to purely empirical research. Despite the example of masters like John Hughlings Jackson, Kurt Goldstein, and James Papez, neurologists consistently avoided any theorizing. Referring to James Conant's analysis in *Modern Science and Modern Man* (Conant 1952), Arieti pointed out that "mental constructions, working hypotheses, are generally frowned upon as armchair speculations, more appropriate to the field of philosophy than to the field of medicine" (Arieti 1956b, p. 324). This prejudicial attitude was, according to him, based on a mistaken conception of modern science, as even in the field of physics "the history of science demonstrate beyond a doubt that the really revolutionary and significant advances come not from empiricism, but from new theories" (Arieti 1956b, p. 324).

The article's rejection reveals that Arieti's positioning on the border between different disciplines and fields of knowledge was neither easy nor comfortable. On one side, his attention toward the biology of human beings was criticized by those who focused on culture and history; on the other, he drew criticism from those scientists who showed in their work the customary "aversion which physicians in particular and biologists in general have for any method which is not strictly empirical" (Arieti 1948, p.325).

Nevertheless, Arieti remained on the border. "Borrowing a popular metaphor from the philosopher Martin Buber," he wrote, "I should say that the student of personality today is like a man walking on a narrow ridge; if he leans too much on one of the two sides, he may fall" (Arieti 1957b, p. 546). In this respect, if his

---

[19] This document is catalogued as undated, but from its content we learn that it dates to March/April 1956, since it refers to the letter sent on March 19, 1956, by Clara Thompson to the Institute members.

[20] In 1956 Silvano Arieti applied for a teaching position at the Institute, and in 1957 he became member of the Faculty of the William Alanson White Institute; in 1958 he also became training and supervising analyst at the Institute.

"scientific humanism"[21] made him one of the most esteemed American psychiatrists, it was not useful for his career, since in "the large psychiatric fraternity" (SAP 1957e) there seemed to be no room for his philosophical and epistemological reflections. So, because of his working on the border Arieti remained on the fringe of academic life. Despite his fame and scientific reputation, he was never offered a chair in an Ivy League university. He continued to work as he thought best and to teach elsewhere.

References

American Psychiatric Association (1969). *New Directions in American Psychiatry, 1944–1968.* American Psychiatric Association.

Arieti, S. (1941). Histopathological Changes in Experimental Metrazol Convulsions in Monkey. *The American Journal of Psychiatry*, 98 (1), 70–76.

Arieti, S. (1942). The Vascularization of Cerebral Neoplasm Studied with the Fuchsin Method of Eros. *The Journal of Neuropathology and Experimental Neurology,* 1 (4), 375–393.

Arieti, S. (1943). Frontal Lobe Tumor Expanding Into the Ventricle. Clinicopathologic Report. *Psychiatric Quarterly,* 17 (2), 227–240.

Arieti, S. (1944a). An Interpretation of the Divergent Outcome of Schizophrenia in Identical Twins. *Psychiatric Quarterly,* 18 (4), 587–599.

Arieti, S. (1944b). The "Placing-Into-Mouth" and Coprophagic Habits Studied from the Point of View of Comparative Developmental Psychology. *The Journal of Nervous and Mental Disease,* 99 (6), 959–964.

Arieti, S. (1944c). Multiple Meningioma and Meningiomas Associated with Other Brain Tumors. *The Journal of Neuropathology and Experimental Neurology,* 3 (3), 255–270.

Arieti, S. (1944d). Progressive Multiform Angiosis. *The Archives of Neurology and Psychiatry,* 51 (2), 182–189

Arieti, S. (1945a). Primitive Habits and Perceptual Alterations in the Terminal Stage of Schizophrenia. *Archives of Neurology and Psychiatry,* 53 (5), 378–384

Arieti, S. (1945b). Primitive Habits in The Preterminal Stage of Schizophrenia. *The Journal of Nervous and Mental Disease,* 102 (4), 367–375.

Arieti, S. (1945c). General Paresis in Senility. Critical Review of the Literature and Clinico-Pathologic Report of Six Cases. *The American Journal of Psychiatry*, 101 (5), 585–593.

Arieti, S. (1946). Histopathologic Changes in Cerebral Malaria and Their Relation to Psychotic Sequel. *The Archives of Neurology and Psychiatry,* 56 (1), 79–104.

Arieti, S. (1947). The Processes of Expectation and Anticipation. Their Genetic Development, Neural Basis and Role in Psychopathology. *The Journal of Nervous and Mental Disease,* 106 (4), 471–481.

Arieti, S. (1948). Special Logic of Schizophrenic and Other Types of Autistic Thought. *Psychiatry,* 11 (4), 325–338.

Arieti, S. (1950a). Autistic Thought. Its Formal Mechanisms and Its Relationship to Schizophrenia. *The Journal of Nervous and Mental Disease,* 111 (4), 288–303.

Arieti, S. (1950b). New Views on the Psychology and Psychopathology of Wit and of The Comic. *Psychiatry,* 13 (1), 43–62.

Arieti, S. (1950c). Primitive Intellectual Mechanisms in Psychopathological Conditions. Study of the Archaic Ego. *The American Journal of Psychotherapy,* 4 (1), 4–15.

[21] This expression about Arieti's work was used in 1976 by David Forrest in a letter sent to *The New York Times Book Review.* This expression will be discussed in Chap. 6.

Arieti, S. (1954). Some Aspects of the Psychopathology of Schizophrenia. *The American Journal of Psychotherapy*, 8 (3), 396–414.
Arieti, S. (1955a). Some Aspects of Language in Schizophrenia. In H. Werner (Ed.), *On Expressive Language* (pp. 53–67). Clark University Press.
Arieti, S. (1955b). *Interpretation of Schizophrenia*. Basic Books.
Arieti, S. (1956a). Some Basic Problems Common to Anthropology and Modern Psychiatry. *American Anthropologist*, 58 (1), 26–39.
Arieti, S. (1956b). The Possibility of Psychosomatic Involvement of the Central Nervous System in Schizophrenia. *The Journal of Nervous and Mental Disease*, 123 (4), 324–333.
Arieti, S. (1957a). What is effective in the Therapeutic Process? A Round Table Discussion. *The American Journal of Psychoanalysis*, 17 (1), 30–33.
Arieti, S. (1957b). The Double Methodology in the Study of Personality and Its Disorders. *The American Journal of Psychotherapy*, 11 (3), 532–547.
Arieti, S. (1957c). The Two Aspects of Schizophrenia. *Psychiatric Quarterly*, 31 (1–4), 403–416.
Arieti, S. (1960). Aspects of Psychoanalytically Oriented Treatment of Schizophrenia. In S.C. Scher & H.R. Davis (Eds.), *The Out-Patient Treatment of Schizophrenia* (pp. 114–118). Grune & Stratton.
Arieti, S. (1967). Rapida rassegna degli studi sul pensiero schizofrenico da Bleuler ai giorni nostri. *Archivio di psicologia neurologia e psichiatria*, 28 (3–4), 237–254.
Arieti, S. (1968). Some Memories and Personal Views. *Contemporary Psychoanalysis*, 5 (1), 85–89.
Arieti, S. (1979a). *The Parnas*. Basic Books.
Arieti, S. (1979b). From Schizophrenia to Creativity. *The American Journal of Psychotherapy*, 33 (4), 490–505.
Balbuena Rivera, F. (2018). "In Honor of Jurgen Ruesch: Remembering His Work in Psychiatry." *International Journal of Social Psychiatry*, 64 (2), 198–203.
Berrios, G., Porter, R. (1995). *A History of Clinical Psychiatry. The Origin & History of Psychiatric Disorders*. The Athlone Press.
Bleuler, E. (1924). *Textbook of Psychiatry*. Macmillan Company.
Bowlby, J. (1979). Psychoanalysis as Art and Science. *International Review of Psychoanalysis*, 6 (1), 3–14.
Buber, M. (1938). *I and Thou*. T. & T. Clark.
Cassirer, E. (1946). *Language and Myth*. Harper & Brothers.
Cassirer, E. (1953). *The Philosophy of Symbolic Forms*. Yale University Press.
Cerea, A. (2015). Il modello della complementarità nelle scienze dell'uomo. Da Niels Bohr a Georges Devereux. *Intersezioni*, 35 (3), 331–354.
Civita, A. (2018). *Psicoanalisi e psichiatria. Storia ed epistemologia*. Edited by A. Molaro. Mimesis.
Conant, J. B. (1952). *Modern Science and Modern Man*. Columbia University Press.
Conci, M. (2008). Gaetano Benedetti in His Correspondence. *International Forum of Psychoanalysis*, 17 (2), 112–129.
Conci, M. (2012). *Sullivan Revisited, Life and Work. Harry Stack Sullivan's Relevance for Contemporary Psychiatry, Psychotherapy and Psychoanalysis*. Tangram Edizioni Scientifiche.
Conci, M., Dazzi, S., & Mantovani M. L. (1997). *La tradizione interpersonale in psichiatria, psicoterapia e psicoanalisi*. Erre emme edizioni.
Dalle Luche, R., Giacobbi, S., & Di Petta, G. (2013). In memoria di Gaetano Benedetti. *Psichiatria e psicoterapia*, 32 (3), 227–233.
Di Petta, G. (2014). Ricordo di Gaetano Benedetti, l'ultimo del Burghölzli. *Comprendre*, 24, 313–315.
Ferraro, A., Arieti, S. (1945). Cerebral Changes in the Course of Pernicious Anemia and Their Relationship to Psychic Symptoms. *The Journal of Neuropathology and Experimental Neurology*, 4 (3), 217–239.
Freeman, W., Watts, J. (1942). *Psychosurgery*. Charles C. Thomas.
Freud, S. (1938). *The Basic Writings of Sigmund Freud*. Edited by A. A. Brill. Modern Library.

Fromm-Reichmann, F. (1950). *Principles of Intensive Psychotherapy.* The University of Chicago Press.
Fromm-Reichmann, F. (1959a). *Psychoanalysis and Psychotherapy.* The University of Chicago Press.
Fromm-Reichmann, F. (1959b). Loneliness. *Psychiatry,* 22 (1), 1–15.
Fromm, E. (1951). *The Forgotten Language.* Rinearth & Co.
Gardner, H. (1985). *The Mind's New Science: A History of the Cognitive Revolution.* Basic Books.
Goldstein, K. (1964). Methodological Approach to the Study of Schizophrenic Thought Disorder. In J. Kasanin (Ed.), *Language and Thought in Schizophrenia Collected Papers Presented at the Meeting of the American Psychiatric Association, May 12, 1939, Chicago – Illinois* (pp. 17–40). The Norton Library.
Goldstein, K., Scheerer, M. (1941). Abstract and Concrete Behavior. An Experimental Study with Special Tests. *Psychological Monographs,* 53 (2), 1–151.
Green, M. R. (1964). *Interpersonal Psychanalysis: The Selected Papers of Clara Thompson.* Basic Books.
Gusdorf, G. (1960). *Introduction aux Sciences Humaines.* Le Belles Lettres.
H.P.P. (1975). Book Notice, Interpretation of Schizophrenia, Second Edition, Completely Revised and Expanded, by Silvano Arieti. *The Bullettin. The New York State District Branches, American Psychiatric Association,* 17 (7–8), 10.
Hilgard, E. R., & Marquis, D. (1940). *Conditioning and Learning.* Appleton.
Hornstein, G. A. (2000). *To Redeem One Person Is to Redeem the World. The Life of Frieda Fromm-Reichmann.* The Free Press.
Jackson, J. H. (1887). An Address on the Psychology of Joking. *British Medical Journal,* 2 (1399), 870–871.
James, W. (1911). *Some Problems of Philosophy.* Longmans, Green & Co.
Jung, C. G. (1936). *The Psychology of Dementia Praecox.* Nervous and Mental Disease Monograph Series.
Kanner, L. (1944). Early Infantile Autism. *The Journal of Pediatrics,* 25 (3), 211–217.
Kanner, L. (1946). Irrelevant and Metaphorical Language in Early Infantile Autism. *The American Journal of Psychiatry,* 103 (2), 242–246.
Kasanin, J. (Ed.). (1964a). *Language and Thought in Schizophrenia. Collected Papers Presented at the Meeting of the American Psychiatric Association, May 12, 1939, Chicago–Illinois.* The Norton Library.
Kasanin, J. (1964b). Introductory Remarks. In J. Kasanin (Ed.), *Language and Thought in Schizophrenia Collected Papers Presented at the Meeting of the American Psychiatric Association, May 12, 1939, Chicago – Illinois* (pp. 1–3). The Norton Library.
Kwawer, J. S. (2017). On Longing and Not Belonging: a Selective History of the William Alanson White Institute. *The American Psychoanalyst,* 51 (2), 7–9.
Langer, S. (1942). *Philosophy in a New Key. A Study in the Symbolism of Reason, Rite and Art.* Harvard University Press.
Levin, M. (1938). Misunderstanding of the Pathogenesis of Schizophrenia Arising from the Concept of Splitting. *The American Journal of Psychiatry,* 94 (4), 877–889.
Lévy-Bruhl, L. (1910). *Les fonctions mentales dans les sociétés inférieures.* Alcan.
Lévy-Bruhl, L. (1922). *La mentalité primitive.* Alcan.
Liebert, R. S. (1956). Book Review. Interpretation of Schizophrenia. By Silvano Arieti. *Psychological Newsletter,* 7, 103–104.
Malmo, R. B. (1942). Interference Factors in Delayed Response in Monkeys After Removal of Frontal Lobes. *Journal of Neurophysiology,* 5 (4), 295–308.
Mann, C. (2001). In Memoriam: Gerard Chrzanowski, MD, 1913–2000. *International Forum of Psychoanalysis,* 10 (1), 94–96.
Mead, G. H. (1934). *Mind, Self and Society.* University of Chicago Press.
Minkowki, E. (1933). *Le temps vécu. Études phénoménologiques et psychopathologiques.* D'Artrey.

Minkowski, E. (1970). *Lived Time: Phenomenological and Psychopathological Studies*. Northwestern University Press.
Minnick, R., Warden, C., & Arieti, S. (1946). The Effects of Sex Hormones on the Copulatory Behavior of Senile White Rats. *Science*, 1904 (2687), 749–750.
Molaro, A. (2013). *Modelli di schizofrenia*. Raffaello Cortina Editore.
Moravia, S. (1972). *Il ragazzo selvaggio dell'Aveyron. Pedagogia e psichiatria nei testi di J. Itard, Ph. Pinel e dell'anonimo della "Décade"*. Laterza.
Moravia, S. (1974). *Il pensiero degli Idéologues. Scienza e filosofia in Francia 1870–1815*. La Nuova Italia.
Moulton, R. (1986). Clara Thompson: Unassuming Leader. In L. Dickstein & C. Nadelson (Eds.), *Women Physicians in Leadership Roles* (pp. 87–93). American Psychiatric Press.
Ortega y Gasset, J. (1936). History as a System. In R. Kilbansky & H. J. Paton (Eds.), *Philosophy and History. Essays Presented to Ernst Cassirer* (pp. 283–322). The Clarendon Press.
Passione, R. (2016a). Fra storia e natura: la psichiatria antiriduzionistica di Silvano Arieti. *Ricerche di psicologia*, 39 (2), 152–179.
Passione, R. (2016b). "La psichiatria di Silvano Arieti: un primo profilo". *Physis. Rivista Internazionale di Storia della Scienza*, 51 (1–2), 219–330.
Passione, R. (2018). Language and Psychiatry: the Contribution of Silvano Arieti Between Biography and Cultural History. *European Yearbook of the History of Psychology*, 4, 11–36.
Passione, R. (2019). "A Glimpse of Inner Struggles". Psychiatry and Human Identity on the Work of Silvano Arieti. *European Yearbook of the History of Psychology*, 5, 83–109.
Pavlov, I. (1941). *Conditioned Reflexes and Psychiatry*. International Publishers.
Pea, R. D. (1982). Werner's Influences on Contemporary Psychology. *Human Development*, 25 (4), 303–308.
Roazen, P., Swerdloff, B. (1995). *Heresy: Sándor Radó and the Psychoanalytic Movement*. Aronson.
Roheim, G. (1934). *The Riddle of the Sphinx*. Hogarth Press.
Ruesch, J. (1957). The Trouble with Psychiatric Research. *The Archives of Neurology and Psychiatry*, 77 (1), 93–107.
Russell, B. (1945). *A History of Western Philosophy*. Simon & Schuster.
Sabshin, M. (1990). Turning Points in Twentieth-Century American Psychiatry. *The American Journal of Psychiatry*, 147 (10), 1267–1274.
Sabshin, M. (2008). *Changing American Psychiatry: A Personal Perspective*. American Psychiatric Publishing.
SAPa. *Correspondence, 1940–1981*.
SAPb. *Speeches and Writings, 1940–1981. Books. Interpretation of Schizophrenia. First edition. Correspondence, 1954–1963*.
SAPc. *Speeches and Writings, 1940–1981. Books. Interpretation of Schizophrenia. First edition. Reviews, 1954–1956*.
SAPd. *Subject File, 1914–1981. Certificate in Psychoanalysis Graduation, 1953*.
SAPe. *Subject File, 1914–1981. Employment. Curriculum vitae, 1940-circa 1980*.
SAPf. *Subject File, 1914–1981. Personal. Divorce, 1964–1965*.
SAPg. *Subject File, 1914–1981. Personal. Photographs, circa 1929–1976*.
SAPh. *Subject File, 1914–1981. Publishing, Italian editors, 1957–1978*.
SAPi. *Subject File, 1914–1981. Publishing. Japanese Translation, 1973–1980*.
SAP (1940). L, May 17 (Elio Arieti to Silvano Arieti). In *Correspondence, 1940–1981*.
SAP (1945a). L, March 2 (Netwon Bigelow to Silvano Arieti). In *Correspondence, 1940–1981*.
SAP (1945b). L, June 5 (William Riegelman to Silvano Arieti). In *Subject File, 1914–1981. Employment. Position appointments, 1941–1961*.
SAP (1945c). L, July 5 (Robert M. Yerkes to Silvano Arieti). In *Correspondence, 1940–1981*.
SAP (1945d). L, August 28 (L.B. Henriques to Silvano Arieti). In *Subject File, 1914–1981. Employment. Position appointments, 1941–1961*.
SAP (1946a). L, January 15 (Charles H. Silver to Silvano Arieti). In *Subject File, 1914–1981. Employment. Position appointments, 1941–1961*.

SAP (1946b). C, January 31 (Certificate of Registration in the Office of the Clerk of the County of New York as Authority to Practice the Profession of Physician and Surgeon). In *Subject File, 1914–1981. Personal. Certificates.*
SAP (1946c). L, March 17 (David de Sola Pool to Silvano Arieti). In *Correspondence, 1940–1981.*
SAP (1946d). F, August 26 (Veterans Administration. Personnel Action). In *Subject File, 1914–1981. Finances. Miscellaneous, 1940–1980.*
SAP (1946e). L, October 7 (Charles F. von Salzen to Silvano Arieti). In *Subject File, 1914–1981. Employment. Position appointments, 1941–1961.*
SAP (1946f). L, November 5 (Howard Potter to Silvano Arieti). In *Subject File, 1914–1981. Employment. Position appointments, 1941–1961.*
SAP. (1946g). L, November 28 (Silvano Arieti to Harry Weinstock). In *Correspondence, 1940–1981.*
SAP (1946h). L, December 5 (US to Silvano Arieti). In *Correspondence, 1940–1981.*
SAP (1946i). L, December 7 (Silvano Arieti to Bettina Warburg). In *Correspondence, 1940–1981.*
SAP (1947a). L, March 12 (Sidney Kahr to Silvano Arieti). In *Correspondence, 1940–1981.*
SAP (1947b). L, April 3 (Martin Steiner to Silvano Arieti). In *Correspondence, 1940–1981.*
SAP (1947c). L, April 7 (Jacob Hetrick to Silvano Arieti). In *Correspondence, 1940–1981.*
SAP (1947d). L, June 13 (Martin Steiner to Silvano Arieti). In *Correspondence, 1940–1981.*
SAP (1947e). L, June 20 (Long Island College Board of Regents to Silvano Arieti). In *Subject File, 1914–1981. Position. Appointments.*
SAP (1947f). TL, July 26 (Elio Arieti to Silvano Arieti)
SAP (1947g). L, 1947g, August 9 (Giulio Arieti to Silvano Arieti). In *Correspondence, 1940–1981.*
SAP (1947h). C, November 28 (County Clerk's Office of Kings County). In *Subject File, 1914–1981. Personal. Certificates.*
SAP (1948a). L, April 26 (Aldo Ballerini to Silvano Arieti). In *Correspondence, 1940–1981.*
SAP (1948b) TL, May 7 (Giampaolo Rocca to Ines Bemporad Arieti). In *Correspondence, 1940–1981.*
SAP. (1948c). TL, May 13 (Elio Arieti to Silvano Arieti).
SAP (1948d). C, May 14 (Department of Health. Certificate of Birth n.18108). In *Subject File, 1914–1981. Personal. Certificates, 1940–1975.*
SAP (1948e). L, August 29 (Aldo Ballerini to Silvano Arieti). In *Correspondence, 1940–1981.*
SAP (1948f). L, September 18 (Silvano Arieti to Harry Stack Sullivan). In *Correspondence, 1940–1981.*
SAP (1948g). L, November 5 (Helen Tepper to Silvano Arieti). In *Correspondence, 1940–1981.*
SAP (1948h). L, December 1 (Edward Bernecker to Silvano Arieti. In *Subject File, 1914–1981. Employment. Position appointments, 1941–1961.*
SAP (1949a). L, February 26 (Eilhard von Domarus to Silvano Arieti). In *Correspondence, 1940–1981.*
SAP (1949b). L, March 14 (Sara Tower to Silvano Arieti). In *Correspondence, 1940–1981.*
SAP (1949c). L, September 19. (Katherine Olinick to Silvano Arieti). In *Correspondence, 1940–1981.*
SAP (1949d). L, October 4 (Angelina della Casa to Silvano Arieti). In *Subject File, 1914–1981. Employment. Position appointments, 1941–1961.*
SAP (1950a). L, February 6 (Helen Tepper to Silvano Arieti). In *Correspondence, 1940–1981.*
SAP (1950b). L, February 14 (Silvano Arieti to Angelina Della Casa). In *Correspondence, 1940–1981.*
SAP (1950c). L, February 27 (Aldo Ballerini to Silvano Arieti). In *Correspondence, 1940–1981.*
SAP (1950d). L, May 23 (Leo Kanner to Silvano Arieti). In *Correspondence, 1940–1981.*
SAP (1950e). PD, 1950e (National Security Resource Board. Health Resources Office. Information Card). In *Correspondence, 1940–1981.*
SAP (1951). L, June 30 (Silvano Arieti to Howard Potter). In *Correspondence, 1940–1981.*
SAP (1952a). L, February 26 (Sandor Rado to Silvano Arieti). In *Correspondence, 1940–1981.*
SAP (1952b). L, May 11 (Silvano Arieti to John Schimel). In *Correspondence, 1940–1981.*
SAP (1952c). L, June 25 (Silvano Arieti to Howard Potter). In *Correspondence, 1940–1981.*

SAP (1952d). L, July 14 (Silvano Arieti to Edward Podolsky). In *Correspondence, 1940–1981.*
SAP (1952e). L, December 18 (Tullio Bazzi to Silvano Arieti). In *Correspondence, 1940–1981.*
SAP (1952f). L, December 30 (Clara Thompson to Silvano Arieti). In *Correspondence, 1940–1981.*
SAP (1953a). L, June 19 (Nicholas Freidberg to Silvano Arieti). In *Correspondence, 1940–1981.*
SAP (1953b). L, July 15 (Heinz Werner to Silvano Arieti). In *Correspondence, 1940–1981.*
SAP (1953c). L, September 15 (Robert Brunner to Silvano Arieti). In *Correspondence, 1940–1981.*
SAP (1953d). L, October 16 (Yolanda Bemporad to Silvano Arieti). In *Correspondence, 1940–1981.*
SAP(1953e). L, December 2 (Eilhard von Domarus to Silvano Arieti). In *Speeches and Writings, 1940–1981. Books. Interpretation of Schizophrenia. First edition. Correspondence, 1954–1963.*
SAP (1953f) . L, December 7 (Robert Brunner to Silvano Arieti). In *Correspondence, 1940–1981.*
SAP (1953g). L, December 9 (Silvano Arieti to Robert Brunner). In *Correspondence, 1940–1981.*
SAP (1954a). L, January 18 (Silvano Arieti to Robert Brunner). In *Correspondence, 1940–1981.*
SAP (1954b). L, May 22 (Clara Thompson to Silvano Arieti). In *Correspondence, 1940–1981.*
SAP (1954c). L, June 2 (Clara Thompson to Frieda Fromm-Reichmann). In *Speeches and Writings, 1940–1981. Books. Interpretation of Schizophrenia. First edition. Correspondence, 1954c–1963.*
SAP (1954d). F, July 6 (American Medical Colleges. The Official Registry of Teaching Personnel). In *Subject File, 1914–1981. Personal. Certificates 1940–1975.*
SAP (1954e). L, October 26 (Al Snow to Silvano Arieti). In *Correspondence, 1940–1981.*
SAP (1955a). L, January 4 (Silvano Arieti to Heinz Werner). In *Correspondence, 1940–1981.*
SAP (1955b). L, February 1 (Arthur Rosenthal to Silvano Arieti). In *Speeches and Writings, 1940–1981. Books. Interpretation of Schizophrenia. First edition. Correspondence, 1954–1963.*
SAP (1955c). L, March 22 (Silvano Arieti to Igino Spadolini). In *Correspondence, 1940–1981.*
SAP (1955d). L, March 25 (Armando Ferraro to Silvano Arieti). In *Correspondence, 1940–1981.*
SAP (1955e). L, May 31 (Silvano Arieti to Heinrich Klüver). In *Correspondence, 1940–1981.*
SAP (1955f). L, May 31 (Silvano Arieti to The American Psychiatric Association). In *Correspondence, 1940–1981.*
SAP (1955g). L, June 9 (Silvano Arieti to Arthur Rosenthal). In *Speeches and Writings, 1940–1981. Books. Interpretation of Schizophrenia. First edition. Correspondence, 1954–1963.*
SAP (1955h). L, June 10 (Arthur Rosenthal to Silvano Arieti). In *Speeches and Writings, 1940–1981. Books. Interpretation of Schizophrenia. First edition. Correspondence, 1954–1963.*
SAP (1955i). L, August 18. (Charles Wahl to Silvano Arieti). In *Speeches and Writings, 1940–1981. Books. Interpretation of Schizophrenia. First edition. Correspondence, 1954–1963.*
SAP (1955j). L, October 31. (Emil Gutheil to Silvano Arieti). In *Correspondence, 1940–1981.*
SAP (1955k). L, November 9 (Barbara Alson to Silvano Arieti). In *Correspondence, 1940–1981.*
SAP (1955l). L, November 18 (Silvano Arieti to Arthur Rosenthal). In *Speeches and Writings, 1940–1981. Books. Interpretation of Schizophrenia. First edition. Correspondence, 1954–1963.*
SAP (1955m). L, November 28 (*Psychiatry* to Silvano Arieti). In *Correspondence, 1940–1981.*
SAP (1956a). L, February 6 (Silvano Arieti to Daniel Blain). In *Correspondence, 1940–1981.*
SAP (1956b). L, February 8 (Russel N. DeJong to Silvano Arieti). In *Correspondence, 1940–1981.*
SAP (1956c). L, February 27. (Roy Grinker to Silvano Arieti). In *Correspondence, 1940–1981.*
SAP (1956d). L, March 8 (Silvano Arieti to Frieda Fromm-Reichmann). In *Speeches and Writings, 1940–1981. Books. Interpretation of Schizophrenia. First edition. Correspondence, 1954–1963.*
SAP (1956e). L, March 19 (Clara Thompson to Graduates and Faculty Members of the William Alanson White Institute). In *Correspondence, 1940–1981.*
SAP (1956f). L, May 8 (Samuel Lazerow to Silvano Arieti). In *Correspondence, 1940–1981.*
SAP (1956g). L, June 9 (Silvano Arieti to Frieda Fromm-Reichmann). In *Correspondence, 1940–1981.*
SAP (1956h). L, October 29 (Leslie Farber to Silvano Arieti). In *Correspondence, 1940–1981.*
SAP (1956i). L, December 10 (Bernard Kaplan to Mabel Blake Cohen). In *Speeches and Writings, 1940–1981. Books. Interpretation of Schizophrenia. First edition. Reviews, 1954–1956i.*
SAP (1956j). L (Silvano Arieti to Windred Overholser). In *Correspondence, 1940–1981.*
SAP (1956k). L (Psychologist graduates to Institute members). In *Correspondence, 1940–1981.*
SAP (1956l). T (*Remarks on the Medical Resolution*). In *Correspondence, 1940–1981.*

SAP (1957a). L, January 29 (Silvano Arieti to Howard Potter). In *Correspondence, 1940–1981.*
SAP (1957b) . L, January 29 (Silvano Arieti to Bernard Kaplan). In *Speeches and Writings, 1940–1981. Books. Interpretation of Schizophrenia. First edition. Reviews, 1954–1956.*
SAP (1957c). L, February 5 (Silvano Arieti to Jurgen Ruesch). In *Correspondence, 1940–1981.*
SAP (1957e). L, December 19 (Lawrence Kolb to Silvano Arieti). In *Correspondence, 1940–1981.*
SAP (1959a). L, February 20 (Howard Potter to Silvano Arieti). In *Correspondence, 1940–1981.*
SAP (1959b). L, March 3 (Silvano Arieti to Howard Potter). In *Correspondence, 1940–1981.*
SAP (1959c). L, October 23 (Paul C. Racamier to Silvano Arieti). In *Subject File, 1914–1981. American HAndbook Of Psychiatry, editor. Correspondence, 1956–1980.*
SAP (1960a). L, March 19 (Bruce Cameron to Silvano Arieti).
SAP (1960b). L, April 9 (Silvano Arieti to Leslie Farber).
SAP (1960c). L, April 16 (Leslie Farber to Silvano Arieti).
SAP (1960d). L, July 27 (Joseph Gabel to Silvano Arieti). In *Correspondence 1940–1981.*
SAP (1961). TMin, 1961 (*Technical Aspects of Terminating Psychotherapy*). In *Speeches and Writings, 1940–1981. Lectures.*
SAP (1968). T, May 11 (Silvano Arieti, *Response to the Presentation of the Frieda Fromm-Reichmann Award*). In *Speeches and Writings, 1940–1981. Lectures.*
SAP (1975). L, January 7 (Silvano Arieti, S. to John Schimel). In *Correspondence, 1940–1981.*
SAP (1977). T, May 25 (Silvano Arieti, *From Schizophrenia to Creativity*). In *Speeches and Writings, 1940–1981. Lectures.*
SAP (1979). T, May 11 (Silvano Arieti, *Cognition in Psychoanalysis*). In *Speeches and Writings, 1940–1981. Lectures.*
SAP (Ua). HN, Unt. (Notes on schizophrenia for *The Intrasychic Self*). In *Speeches and Writings, 1940–1981. The Intrapsychic Self: Feeling, Cognition and Creativity in Health and Mental Illness. Notes, circa 1966.*
SAP (Ub). L (Bernard Kaplan to *Psychiatry*). In *Speeches and Writings, 1940–1981. Books. Interpretation of Schizophrenia. First edition. Reviews, 1954–1956.*
SAP (Uc). L (Silvano Arieti to UR).
SAP (Ud). PD (Arieti's business card). In *Subject File, 1914–1981. Personal. Memento.*
SAP (Ue). T, (Gerard Chrzanowski, *Profile: Silvano Arieti*). In *Subject File, 1914–1981. Personal. Profile, undated.*
SAP (Uf). T (Silvano Arieti, *Outline of Course on Schizophrenia*). In *Subject File, 1914–1981. Teaching. Course on Schizophrenia.*
SAP (Ug). T (Silvano Arieti, *Preface*). In *Speeches and Writings, 1940–1981. Books. Interpretation of Schizophrenia. First edition. Publishing, 1954–1955.*
Schaffer, L. (1956). Interpretation of Schizophrenia. By Silvano Arieti. *Psychiatry,* 19 (3), 319–323.
Shapiro, S. A. (2017). The History of the William Alanson White Institute Sixty Years After Thompson. *Contemporary Psychoanalysis,* 53 (1), 44–62.
Spiegelberg, H. (1972). *Phenomenology in Psychology and Psychiatry. A Historical Introduction.* Northwestern University Press.
Stern, D. (1995). *Pioneers of Interpersonal Psychoanalysis.* Analytic Press.
Storch, A. (1924). *The Primitive Archaic Forms of Inner Experiences and Thought in Schizophrenia.* The Nervous and Mental Disease Publishing Company.
Sullivan, H. S. (1955). *Conceptions of Modern Psychiatry. The First William Alanson White Memorial Lectures.* Tavistock Publications.
Sullivan, H. S. (1964). The Language of Schizophrenia. In J. Kasanin (Ed.), *Language and Thought in Schizophrenia Collected Papers Presented at the Meeting of the American Psychiatric Association, May 12, 1939, Chicago – Illinois* (pp. 4–16). The Norton Library.
Thompson, C. (2017). The History of the William Alanson White Institute. *Contemporary Psychoanalysis,* 53 (1), 7–28.
Tomlinson, C. W. (2010). Sandor Rado and the Fate of the Berlin Model in New York. *Psychoanalysis and History,* 12 (1), 97–107.

Villeneuve, A. (1965). Influence of Phenomenologic and Existential Philosophies on Contemporary Psychiatry. *Psychiatric Quarterly,* 39 (1–4), 91–101.

von Domarus, E. (1964). The Specific Laws of Logic in Schizophrenia. In J. Kasanin (Ed.), *Language and Thought in Schizophrenia Collected Papers Presented at the Meeting of the American Psychiatric Association, May 12, 1939, Chicago–Illinois* (pp. 104–114.) The Norton Library.

Vygotskij, L. S. (1934). Thought in Schizophrenia. *The Archives of Neurology and Psychiatry,* 31 (5), 1062–1077.

Wapner, S., & Kaplan, B. (1964). Heinz Werner: 1890–1964. *The American Journal of Psychology,* 77 (3), 513–517.

Ward, M. J. (1946). *The Snake Pit.* Random House.

Werner, H. (1940). *Comparative Psychology of Mental Development.* Harper & Brothers.

Windholz, E. (1947). In Memoriam. Jacob S. Kasanin 1897–1946. *Psychoanalytic Quarterly,* 16 (1), 94–95.

Yerkes, R. M., Yerkes, D. N. (1928). Concerning Memory in the Chimpanzee. *Journal of Comparative Psychology,* 8 (3), 237–271.

Chapter 4
Roots

4.1 Integration

In the late 1950s, American psychiatry was increasingly characterized by an acerbic harsh confrontation of different schools (Sabshin 1990, 2008). When, in 1956, the William Alanson White Institute requested accreditation from the American Psychoanalytic Association, the request was rejected, most likely as a result of the theoretical independence of the Institute, as exemplified in its disagreement with Freudian orthodoxy. In response, a group of "heretical" psychiatrists and analysts gathered to found the American Academy of Psychoanalysis. The group included Franz Alexander, Gerard Chrzanowski, Frieda Fromm-Reichmann, Janet Rioch, Rose Spiegel, Clara Thompson, as well as Sandor Rado (Burhnam 1976; Conci 2012), whose work to counteract narrow views Silvano Arieti greatly valued.

Arieti, whose independent and original voice focused on psychiatry as a discipline at the boundary between natural and human science, on October 5, 1956, submitted his request to become a member of the Academy, for he wholeheartedly supported its cause, opposing the view that psychoanalytic training must subscribe to a closed-minded unitary orthodoxy (SAP 1956l, 1956n).

Arieti personally experienced this tense climate, when he learned from David Engelhardt, Assistant Director of the Department of Psychiatry at New York Medical College, that the Department planned to cancel his course. He immediately wrote:

> Dear Doctor Engelhardt, I was surprised to hear from you yesterday on the telephone that you intend to abolish my course. I know that the residents are very busy, that they already have a course on the psychoses, and that there is some opposition to my course on theoretical grounds. I feel, however, that in spite of this opposition young psychiatrists should not be exposed exclusively to the orthodox psychoanalytic approach. If early in their career they are not confronted with other psychoanalytic and psychiatric points of view, it will be much more difficult to enlarge their horizons later (SAP 1956m).

Caught between orthodoxies—he was both critical of neurologists and too eclectic for psychiatrists—Arieti asked Engelhardt to change his mind, and in the letter he increased the stakes with his projected new work: "Basic Books will publish an

R. Passione, *Psychiatry and the Human Condition*, Springer Biographies,
https://doi.org/10.1007/978-3-031-09304-3_4

American Handbook of Psychiatry and has invited me to be the editor. Many people from various schools will participate ... This book ... have the purpose of integrating the various schools in the large body of psychiatric knowledge" (SAP 1956m).

It was not, in fact, Basic Book's initiative but his own, and his idea was for a *Textbook*[1] (SAP 1956h) that integrated different perspectives rather than demarcating boundaries. His goal was to reduce the contentious climate of American psychiatry with dialogue and integration (Arieti 1959, 1966).

Arieti was aware, of course, of the innovation of his enterprise. Whereas Europe was the land of collective textbooks, whose homogeneity and consistency were ensured by a general adherence to a neuropsychiatric orientation, American psychiatry lacked such a tradition, owing to the existence of many contrasting schools of thought. How could they all be represented in the same book without compromising its coherence? Answering this question implied a choice between consistency and pluralism, and Arieti definitely chose pluralism. "I felt that we psychiatrists must accept reality even in the field of psychiatry," he stated with sharp self-irony. "The present reality is of a kind where uniformity and consistency are lacking. The beginner must know this truth, even if it is unpleasant ... An obstinate search for consistency at the present stage of our knowledge has so far led some of us not to major syntheses, but to reductionisms of various sorts" (SAP Uf).[2]

In June 1956 Arieti wrote to Arthur Rosenthal, President of Basic Books, to describe his project. It was not to be a compendium of papers, but a book of original articles by leading figures in the different specialized fields of American psychiatry. "I think that this book could be a great contribution to science and would reward the publishing firm both in terms of money and prestige" (SAP 1956d). As a forward-looking and enterprising publisher, Rosenthal welcomed Arieti's request and invested a sum of one hundred and fifty thousand dollars on the book. The contract was signed in September 1956. "All of us here look forward with much enthusiasm to working closely with you on this exciting project," Rosenthal wrote, assuring him that the publishing firm "shall assist the editor in every way"—i.e., over and above the copy editing, and especially in the work's promotion (SAP 1956k).

In this way the *Handbook* became a prominent volume in the planning of Basic Books. Rosenthal believed that high-quality, groundbreaking projects were always rewarding. "We are being well repaid for concentrating in areas where we believed we could make a contribution to American thought and culture," he observed, explicitly referring to Arieti's work, during a meeting with the presidential candidate John Fitzgerald Kennedy (Steinberg 1960, p. 41). "It was nice to see your picture with Kennedy and to see the *Handbook* mentioned in the article," Arieti wrote, after reading the report in *The New York Herald Tribune* (SAP 1960f).

---

[1] This is the word initially used. Later this term was abandoned and replaced with *Handbook*.

[2] These notes were probably prepared for a public presentation of the *Handbook*. The incipit read: "Perhaps October 1959 will be a date which will be remembered. The first American textbook of psychiatry, a collective effort by and large representing psychiatric knowledge in this country, saw the light of day".

The first two volumes of the *American Handbook of Psychiatry* appeared in 1959. Soon afterwards, a new project took shape—a third volume on recent developments in biochemical, community, and conceptual (or cognitive) psychiatry. In 1963 the contract was drawn up, and three years later the new volume was published (SAP 1963b; Arieti 1966).

The *Handbook* brought together more than one hundred American psychiatrists. Arieti chose the members of the Editorial Board with scientific and strategic intelligence, relying on acknowledged authorities: Kurt Goldstein, for his high and undisputed scientific reputation; Lawrence Kolb, Professor of Psychiatry at Columbia University and Director of the New York State Psychiatric Institute; Norman Cameron, Professor of Psychiatry at Yale; Daniel Blain, Medical Director of the American Psychiatric Association (SAP 1956f, 1956g, 1956i); and Kenneth Appel, a consultant for legislation and for the relationship between the psychiatric community and the U.S. Congress (Barton 1987).

Silvano Arieti compared the assembly of the *American Handbook* to the construction of a cathedral, with himself as the architect. He directed the project with diplomatic skill and a firm hand, organizing everything meticulously and managing relationships with the authors sensitively and frankly (SAPg). Scientific conversations with the members of the Editorial Board were often intense; though discussions covered all relevant issues concerning the contents and the quality of contributions, Arieti was responsible for the final decisions. The case of Carl Rogers and his chapter on client-centered-therapy is illustrative. At the beginning of 1957, Harry Bone (a contributor who was a member of the group of psychologists whose training at the William Alanson White Institute had been debated shortly before) (Chap. 3; SAP 1956c) had suggested its inclusion in the *Handbook*: "I think a big gap in the presentation of contemporary psychotherapy would appear in our book if this therapy is not included" (SAP 1957a).More particularly, Bone emphasized that the inclusion of this "native American non-medical therapy" could help to increase communication and mutual understanding between medical and psychological therapists. Despite Arieti's initial hesitation—since Rogers' approach did not seem entirely convincing to him—Bone's call to dialogue eventually persuaded him (SAP 1963c), notwithstanding the contrary opinion expressed by the members of the Editorial Board, who considered the client-centered-therapy as an "offbeat and naive method" (SAP 1957b, 1957c).

Silvano Arieti therefore reaffirmed pluralism as the *Handbook*'s guiding principle and ensured its application within each section of the book. Another example of Arieti's commitment to pluralism is his correspondence with Joseph Cramer about the chapter on childhood neuroses:

> Dear Dr. Cramer, I wish to thank you for sending us your manuscript which I myself have immediately reviewed ... I consider your paper an excellent piece of work ... Nevertheless, I wish to express some comments to you.
>
> Although each author is entitled to select his own theoretical frame of reference, the main alternative points of view should be mentioned in a text-book. You have overlooked this recommendation made by the Editorial Board ... Have you taken for granted that everybody will accept the Freudian frame of reference? What about other points of view which focus on

the total interpersonal parent-child relationship? ... It seems to me that you have somewhat neglected the social relations (SAP 1957f).

Arieti's notes to Hervey Cleckley about his chapter on psychopathic states conveyed the same import, since Cleckley had not considered psychodynamic interpretations: "It is the policy of the Editorial Board to respect the points of view of the authors, and certainly we do not want to impose a psychoanalytic frame of reference on those who are not psychoanalytically inclined. However, we feel that in a text-book of this kind, in addition to his own, the author must expound the current alternative points of view" (SAP 1957e).

Arieti's comments were not a partisan warning coming from a dynamic psychiatrist, but a carrying out of the principle that a pluralistic approach be respected throughout the book. According to this principle, for example, he paid particular attention to the chapters on organic conditions, exhorting the authors to give full scope to the presentation of neuropathological data (SAP 1957d). In short, he consistently ensured that the *Handbook* not neglect the neuropsychiatric school of thought: "Certainly we must add new dimensions, but not at the expenses of the traditional ones, which still constitute the backbone of our science" (SAP 1969t).

Work on the *Handbook* was long and hard. Following every step, discussing the details, coordinating the authors and goading them to meet deadlines were not easy tasks. Here Arieti began to experience a host of worries, the number of which would multiply with the *Handbook*'s second edition, enlarged to seven volumes (Arieti 1974–1975; Arieti & Brodie 1981). Difficulties aside, however, by means of this work Arieti was developing deep roots in American psychiatry, both *integrating it*, in accordance with the work's original plan, and *integrating himself* into it. Once again, science and biography were woven together, joined by the "dialogic principle" that characterized Arieti's life and work.

Publication was a great success. Nothing like the *Handbook* had ever been attempted for American psychiatry. It was repeatedly heralded in its many American and foreign reviews for its newness and scope. It was hailed as a "Herculean task" (JIMP 1960, p. 2987), and the "risk of pluralism" that Arieti had decided to take was fully approved by Donald Hastings in *Science* (Hastings 1960). It was almost a book *on* psychiatry rather than *of* psychiatry, suggested another reviewer, describing it as "the most monumental work that has appeared upon the psychiatric scene" (Kilpatrick 1961, p. 292). Above all, by putting together many different and even contrasting perspectives, the book demonstrated that "we cannot take anything for granted" (DeRosis 1959, p. 3048) and that reputedly stable and definitive views must continuously give way to new data and evidence. In this way, the *Handbook* was an original contribution to the debate on the scientific status of psychiatry, since it showed that psychiatry, like any other "scientifically determined theory, had the characteristic of being subject to change" (DeRosis 1959, p. 3048).

The *Handbook* increased Arieti's fame, as shown by his bountiful scientific correspondence and by letters he received from all over the country. This success did not please all, however. The Menninger Foundation, for example, complained that it was barely represented in the book. "The general tone of this work is distinctly on the conservative side," read a brief review that appeared in the "Reading Notes" of the Clinic's *Bullettin*, edited personally by Karl Menninger. "One author concedes that the Veterans Administration and some large institutions such as The Menninger Foundation continue to maintain an extensive program of training. Other than this, we are not mentioned in the two volumes" (Menninger 1961, p. 266). This bitter remark may have been prompted by some regrets, for Menninger had actually declined Arieti's invitation to contribute to the work (SAP 1956j).

In 1961 Silvano Arieti resigned his teaching position at the State University and moved to New York Medical College (SAP 1961c, 1961d, 1961e), where a Comprehensive Course of Psychoanalysis had been established in 1944 (Pérez, Moskowitz & Javier 2004). The same school that had rejected his application as a teacher in 1947 now opened its doors to him. Besides teaching, here Arieti dealt with the supervision of residents at Flower Fifth Avenue Hospital in Manhattan. In 1964, owing to his many commitments, he asked for a temporary leave of absence from this appointment. In addition to the third volume of the *Handbook*, he was working assiduously on another book that took much of his time; and—last but not least—he had been elected President of the William Alanson White Psychoanalytic Society, another time-consuming job. "I am sure that if I am alive and in good health I could start doing supervision by September 1965," he wrote to Alfred Freedman, Director of the Department of Psychiatry at Flower Fifth Avenue Hospital (SAP 1964e). He kept his promise, resuming supervision at the Hospital in 1965.

Invitations to conferences, lectures, seminars, and training workshops continued to follow one after another during the 1960s. His other obligations included psychotherapy with patients at Gracie Square Hospital, in Manhattan, and at High Point Hospital, in Port Chester, New York; and with the patients of his own private practice. At the same time, he was continuing his extensive personal scientific correspondence, as he never failed to reply to the many queries and requests from young students for advice and discussion (SAPa). "He liked to discuss ideas with developing professional and would spend innumerable hours reading and correcting their manuscripts and then take entire evenings to minutely go over the junior persons' work in progress" (Bemporad 1981, pp. vi–vii). These words, written by Jules Bemporad, find full confirmation in the analysis of Arieti's correspondence.

Thus, at the age of fifty Silvano Arieti had achieved a great scientific success. The fame that he had achieved with *Interpretation of Schizophrenia* was significantly strengthened and expanded.

Fame and success plunged him into the core of American history in the matter of the assassination of President John Fitzgerald Kennedy when he was called to be a consultant in the case against Jack Ruby. On December 13, 1963, Arieti was telephoned by the prosecuting attorneys, Henry Wade and William F. Alexander, who were heeding the advice of their psychiatrist expert, John Holbrook, to ask Arieti to join them in Texas to evaluate the mental state of Jack Ruby, who on November 24 had shot and killed Lee Harvey Oswald in the basement of the Dallas police headquarters (SAP 1963f).

Holbrook believed that at the time of the murder Ruby was able to distinguish right from wrong and to understand the nature and consequences of his act. To convince the court to accept his opinion, anyway, he had to refute the defense team headed by Melvin Belli, reputed to be a "king of the courtroom" (Shaw 2007). To exonerate the accused or at least to spare him the electric chair, Belli gambled on Karl Menninger's concept of "episodic dyscontrol," a consequence of an epileptic constitution and emotional instability (Menninger & Mayman 1955; Satten, Menninger, Rosen & Mayman 1960). For this reason, Holbrook and the prosecuting attorneys decided to call on a high-profile psychiatrist like Arieti, who answered the call and promptly flew to Dallas (SAPk).

Because Ruby's lawyers refused any further examination of their client by the State's physicians, Arieti could not see him immediately and had to rely on Holbrook's notes on Ruby's mental state (SAP 1963e, 1964a). However, it seems that Arieti eventually succeeded in visiting Ruby, as indicated by some handwritten notes taken during the examination: "I am not going to talk to you. I leave or you leave. Lucid, clean. Extrovert. Loves publicity … No interests in politics … No episodic dyscontrol. Previous actions focused at implementing of crime. Awareness of what he did at the time of crime … He wanted to prevent Mrs. Kennedy from coming to the trial" (SAP Ua).

The first round of Ruby's trial ended on March 11, 1964, when all the prosecution's experts were suddenly called to testify. Given the unexpected timing, Silvano Arieti could not be physically present; nevertheless, Holbrook, in his speech to the Court, brought up the name of Arieti, reporting his evaluation of the accused (SAP 1964b, 1964c). Ruby was ultimately sentenced to death, but his case did not end there. On March 23, 1964, Holbrook wrote to Arieti from Dallas, "It is my feeling that you will ultimately be called as a witness before litigation is completed in this matter" (SAP 1964b). Still, the trial would go in a different direction, and Arieti would not be called.

In this whole affair Arieti remained behind the curtain. Despite the journalistic bedlam, which often brought psychiatrists' voices into the foreground, he never spoke in public. He participated in the Ruby case as a consequence of his established fame and not as a means to further prominence. His interest was purely of a scientific and intellectual kind, as the debate on Ruby's culpability involved a discussion of a human being's voluntary action and capacity to choose by an independent free will—issues that Arieti was to discuss a few years later in another book, *The Will To Be Human* (Arieti 1972).

The experience in Dallas shows another facet of Arieti's rootedness in American culture and history—a rootedness that he also developed at an institutional level, by participating in the life of the organizations to which he belonged. In those years, his position within the American Psychiatric Association was enhanced (SAP 1961f, 1965e, 1965f; SAP 1967a, 1967b),[3] as it was in the American Academy of Psychoanalysis, where, as a member of the Research Committee, he worked to establish a

[3] In 1961 Arieti became a member of the Publications Review Committee, of which in 1967 he was appointed chair. Furthermore, in 1965 he also became a member of the Editorial Board of *Psychiatric News. Official Newspaper of the American Psychiatric Association.*

closer relationship with the American Association for the Advancement of Science (SAP 1964f, 1964g; Ullman 1965).

At the same time, Arieti began to collaborate with the National Institute of Mental Health, where a special Research Committee on schizophrenia had been established (SAP 1966e, 1966f, 1966g) and where in November 1967, before "a record audience of about 500 people" (SAP 1967g), he delivered his Frieda Fromm-Reichmann Memorial Lecture. The lecture occurred six months before he received, on May 11, 1968, the Frieda Fromm-Reichmann Award for his "extraordinary achievements in the diagnosis and treatment of schizophrenia" (SAP 1969f). On receiving the award, Arieti gave an emotional speech in which he reviewed his life, pointing out two meaningful steps of his journey: first, being torn from his home and family and his escape from Italy on a cold January night; and second, his warm embrace by the audience as he was on the stage receiving the Frieda Fromm-Reichmann Award. His story seemed to prove that making one's way might be difficult, but was not impossible (SAP 1968f; Arieti 1968a).

4.2 New Lands to Cultivate

Despite being deeply entrenched in the multifaceted psychiatry of mid-century America, Silvano Arieti was also markedly unconventional in his thinking, for he paid more attention to cognition than was the case for most psychiatrists, who typically tried "to solve everything by recourse to the emotional life" (SAP 1953, my translation). This distinctive feature of his work had emerged in his early observations of perceptual alterations in schizophrenic patients and had prompted him to reconsider Gestalt theories, since schizophrenics showed a pronounced fragmentation of perception and seemed uncapable of perceiving an object as a whole (Arieti 1945, 1965).

Aimed at a structural analysis of thought, this original cognitive line of research was later combined with the dynamic approach in his articles "The Process of Expectation and Anticipation" and "Special Logic," which had thrust Arieti into the limelight of American psychiatry (Arieti 1947, 1948). These works, however, did not generate a new school of thought in America in the 1940s and 1950s, whereas in France, owing to the work of Lévi-Strauss, the structural approach was becoming known.

"Psychology, under the influence of behaviorism, was concerned mainly with overt behavior, classic psychoanalysis focused on the energetics and instinctual precognitive life, and neo-Freudian, cultural psychoanalysis was concerned with the study of conflicts without considering their cognitive origins" (SAP 1979). In 1957 Noam Chomsky paved the way for cognition in the field of linguistics (Chomsky 1957); but it took some time before the echoes of his insights had a wide impact on other disciplines. Thus, until the end of the 1960s, cognition remained "the Cinderella of psychoanalysis and psychiatry. No other field—wrote Arieti—has been so consistently neglected by clinicians and theoreticians alike" (Arieti 1965, p. 16).

It seems paradoxical that in the country where the idealism of the founding fathers was a major substrate of American history, there was "almost a sense of prudish embarrassment in admitting that ideas count" or in investigating deeply "what ideas do to men, and what men do to ideas" (Arieti 1977, p. 6). This reluctance to acknowledge ideas as actual components of the psyche was still more incomprehensible in therapeutic practice: "I do not need to point out to anyone that in psychoanalytic therapy we deal with ideas constantly, and that almost all our exchanges with patients occur through ideas, and that it is through ideas that we bring about improvement or cures" (Arieti 1977, p. 6), Arieti wrote. In 1979 he remarked:

> If some colleagues who have not been interested in this subject were to ask me to define cognition, and hear my answer, they would feel like Moliere's famous character, Mr. Jourdain, who, when his teacher explained what prose was, said *I have spoken prose all my life without even knowing it.* Similarly, even those of us ... who have not been interested in cognition have done cognitive psychoanalysis every day, during every session, because cognition is the study of ideas and their precursors, that is, the study of the development, formation, content, interconnections and dynamic effect of ideas. It is through ideas that we communicate with our patients; it is by hearing the content of their ideas that we get to know them and to know what ideas do to them (SAP 1979).

This emphasis on ideas and cognition did not entail minimizing the importance of affective life and its role in human motivation; if anything, paying attention to ideas acknowledged that most human emotions would not exist without a cognitive basis. As Arieti suggested, one need consider only the evidence provided by Papez's research showing the role played by neocortical areas in emotional states (Arieti 1977).

According to Arieti, the general resistance to the study of cognition revealed a feature typical of American culture, where the role of foundational ideas (such as those of freedom, equality, independence, and justice) gave rise to the belief that human beings are the result of the molding influences of the external world—a view attributable to the impact of John Locke on American culture, as Gordon Allport pointed out in 1955 (Allport 1955; Arieti 1963a).

Things were to change. Many contributions brought about the change, just as many tributaries flow together to form a river. In 1956, a year before Noam Chomsky's groundbreaking *Syntactic Structures*, the Symposium on Information Theory, held in Boston at the Massachusetts Institute of Technology, had paved the way for a new attention to cognition—a turning point for psychological research (Gardner 1985). Then, in 1963, John Flavell published *The Developmental Psychology of Jean Piaget*, which brought the work of the Swiss psychologist to the attention of an American audience (Flavell 1963). Arieti welcomed these contributions. He had been working on cognition for years, and, he said, while this wait had been taxing, it had not been "a waiting for Godot" (SAP 1979), for cognition was finally becoming a subject of study.

Also in 1963, a chapter on conceptual or cognitive psychiatry took shape in the third volume of the *American Handbook*:

> Conceptual, or cognitive, psychiatry is still in a developing stage ... It emphasizes, first of all, that conceptual life as a dynamic force has been greatly underestimated. Up to now the stress

> has been on the primitive, on bodily needs, instinctual behavior, and primitive emotions which can exist without a cognitive counterpart or with a very limited one. Conceptual psychiatry does not deny that lack of gratification of primitive needs may unchain powerful dynamic forces and lead to conflicts. It asserts, however, that emotional factors ... are by no means all derivatives of this type of deprivation. In the middle of the twentieth century fewer men than ever before starve for food or sex; yet psychological malaise and mental illness have increased, not decreased. Conceptual life is the source and sustenance of powerful psychodynamic forces (Arieti 1966, p. x).

Thus, the time had come to acknowledge that pathogenic conditions have to do more with ideas, representations, self-image, and world-view, than with primary needs.

There was still a long way to go, however, starting within the Neo-Freudian school, whose adherence to Locke's conception of man had led to a neglect of the intrapsychic sphere. At the William Alanson White Institute only a few members had dared to venture in this direction along three different lines of research: the study of cognition from a developmental point of view (Ernest Schactel and Arieti himself), the analysis of prelogical experience (Edward Tauber and Maurice Green), and the investigation of communication (Rose Spiegel, Edwin Weinstein, Joseph Jaffe) (Arieti 1965).

In 1964, upon being elected President of the William Alanson White Psychoanalytic Society, Silvano Arieti worked to strengthen this line of research, with the aim of promoting a "cognitive renewal" of the school. His presidential address focused on innovation. It was time, he said, to emend the merely external point of view that asserted a conception of humans as a *tabula rasa.* Arieti's message was not about returning to the orthodox position focused on the Freudian concept of inner drives: it was about recognizing the relevance of the psyche's other dimensions and other internal factors, such as those pertaining to the intrapsychic and cognitive spheres (SAP 1964d).

In his speech Arieti distanced himself not only from behaviorism, which described a human being as a mere *machine* activated by conditioning, but also from the views of the most radical neo-Freudians, who described a human as molded only by cultural and social influences. In contrast, Silvano Arieti's conception of man was inspired by the work of the Austrian biologist Ludwig von Bertalanffy, who argued that living organisms are endowed with an activity of a spontaneous kind that allows them "to transcend the usual psychological formula stimulus–response" (SAP 1964d).

Von Bertalanffy, whom Arieti called into play here for the first time, became in those years a key figure in the theoretical renewal of life sciences and human sciences (von Bertalanffy 1964, 1968a, 1968b, 1969, 1981). Professor of Theoretical Biology at the University of Alberta, Canada, von Bertalanffy was the father of a new "science of systems," a broad conceptual reorientation that took place at the same time in biology, social sciences, applied fields, and even in physics, under the banner of notions such as wholeness, organization, goal-directedness, regulation, complexity, and anti-reductionism (Lazlo 1972; Taschdjian 1975; Davidson 1983). According to von Bertalanffy at the UNESCO in 1965:

> General systems theory is intended to elaborate properties, principles and laws that are characteristic of 'systems' in general, irrespective of their particular kind, the nature of their

> component elements, and the relation of 'forces' between them. A 'system' is defined as a complex of elements in interaction, these interactions being of an ordered (non-random) nature. Being concerned with formal characteristics of entities called systems, general systems theory is interdisciplinary, that is, can be employed for phenomena investigated in different traditional branches of scientific research. It is not limited to material systems but applies to any whole consisting in interacting components (von Bertalanffy 1965).

In the 1960s General Systems Theory (GST) became the keystone of a multifaceted knowledge, the rich articulations of which mirrored the complexity of the human being. Roy Grinker called GST the third modern psychiatric revolution (after psychoanalysis and behaviorism) (Grinker 1969, 1976).

With its interdisciplinarity, GST represented a frame of reference of that composite, seemingly fragmented, multifaceted, pluralistic and sometimes even inconsistent knowledge that Arieti put on display in the *Handbook*, to the third volume of which he invited von Bertalanffy to contribute (SAP 1963d). In addition, in the late 1960s Silvano Arieti actively participated in the special program on GST scheduled by the American Psychiatric Association (SAP 1966b, 1966d, 1966h, 1967c, 1967d, 1969p, 1969q, 1974a, 1974b).

Thus, GST seemed an ideal theoretical counterpart of the methodological approach that was the basis of Arieti's work. But also, above all, Arieti shared with von Bertalanffy another key-idea—that the mind is not a passive recipient. Upon reading the manuscript of *General System Theory*, in 1968 Arieti wrote:

> Dear Ludwig, ... the reading ... was a real fest [*sic*][4] to the spirit and to the intellect. My reaction was enthusiastic... First of all, because I learned from it many things which I did not know; secondly, because the book clarified other things which I already knew, but not in such a clear and structured manner. In the third place, I saw in the book, *si parva magnis licet comparare*, a reproduction or, to be more exact, an antecedent of the battle that I, myself, have had to wage for the last few decades in the field of psychology and psychiatry. I, too, ... have been adverse to conceive the mind as a passive recipient. I, too, ... have emphasized the role of cognition ... I, too, ... had troubles with those who, outSullivaning Sullivan, consider the state of mental health and pathological conditions ... exclusively as reactions to environmental forces (SAP 1968c).

It is here that we can understand the reason for Arieti's reference to von Bertalanffy in his presidential address at the William Alanson White Institute, where he knew that there were many *outSullivaning Sullivan* colleagues to counteract and convince. His speech at the Institute was aimed at promoting a real change and therefore might have sounded almost provocative, so much so that Arieti decided to soften its tone for publication (SAP 1964d; Arieti 1964). Still, he later returned to the matter. For example, upon being awarded the Gutheil Memorial Medal by the Association for the Advancement of Psychotherapy, he criticized the Lockean conception of man as *tabula rasa,* and urged dynamic psychiatry to include analysis of the intrapsychic. In his speech at the award ceremony he said:

> My training at the White Institute taught me to study the world which the child meets and the child's way of experiencing that world, especially in the interpersonal aspects. It

[4] Most probably this Italianism was not only a typographical error, but a superimposition of the Italian word "festa" on the English word "feast".

> also taught me what people can do to one another ... I was gradually filling the numerous gaps and doubts that my original observations at Pilgrim State Hospital had left in me. All this for the good. But soon other doubts started to creep in, and I saw different types of gaps. By stressing the interpersonal, Sullivan and the interpersonal school did not intend to subtract the intrapsychic, but in practice many Sullivanians did so ... The inner self was neglected, at least in theoretical conceptions. It is true that inasmuch as the human being is strongly influenced by the environment ... we must acknowledge in him a fundamental state of *receptivity*. But I felt that he cannot be defined only in terms of a state of receptivity. Every human being, even in early childhood, has another basic function which we can call *integrative activity*. Just as the transactions with the world not only inform but transform the individual, with his integrative activity the individual transforms these transactions and in his turn he is informed and transformed by these transformations. No influence is received like a direct and immutable message. Multiple processes involving interpersonal and intrapsychic dimensions go back and forth (Arieti 1979, p. 493).

In his study of the intrapsychic Arieti investigated a vast range of scientific literature, in which old and new landmark ideas mingled—from Piaget to von Bertalanffy; from the unconscious cognition of the Würzburg school (Narziss Kaspar Ach, Karl Marbe) to Heinz Werner's conception of microgeny, which Arieti used as a key concept to understanding normal and pathological thinking:

> Microgeny, as illustrated by Werner, is the immediate unfolding of a phenomenon, that is, the sequence of the necessary steps inherent in the occurrence of a psychological process. For instance, to the question *Who is the author of Hamlet?* a person answers *Shakespeare*. He is aware only of the question (stimulus) and of his answer (conscious response), but not of the numerous steps which in a remarkably short time led him to give the correct answer ... The numerous steps which a mental process goes through constitute its microgenetic development ... When a process goes through a sequence of stages and forms, as they usually evolve in the average human being, we have a normal psychological process. When sequences of stages and forms combine to produce harmful or undesirable effects we have the psychopathological process. When sequences of stages and forms combine to produce something new and useful we have the creative process (Arieti 1965, pp. 24-25).

In his analysis of thought, Arieti suggested a sort of dissection: his approach was aimed at dividing mental processes into their basic components. In this respect, at the basis of Arieti's theoretical thinking were atomism and elementism, as demonstrated also by his explicit reference to Wilhelm Wundt's elementistic conception of mind: "The major psychological functions with which man is less or more provided at birth may be divided into the three fundamental categories of cognition, affect, conation. This classification, adhered to by Wundt, has always been under attack by those who see in it a perpetuation of the atomistic or associationistic view of the psyche. Actually this classification not only has worthwhile descriptive and deeply-rooted conceptual values, but *grosso modo* corresponds to different anatomical structures in the central nervous system" (SAP Ug).

This approach may suggest a direct influence of the classical cognitive science that was developing in those years on the wave of information theory, the impact of which would find confirmation in Arieti's definition of thinking as a "process by which past experiences are properly stored, organized, recalled, reorganized in new ways" (Arieti 1962a, p. 455). Nevertheless, despite his positive opinion about the advent of information theory as a revitalizing and innovative approach in psychological research (SAP 1968b), Arieti did not agree with other relevant key concepts of the contemporary cognitive revolution, the most important of which was the analogy of the mind to a computer. "Thinking does not consist only of searching for data," he wrote in 1962, emphasizing his distance from the computational conception of cognition (Arieti 1962a, p. 455). He later remarked this distance when he described the approach of classical cognitive science as a new form of positivism (Arieti 1969a, 1971). What the conception of the mind as a computer left behind was the role played by subjectivity in psychological processes. For Arieti, cognition was not a simple system for transmitting and processing data, but a nexus of thoughts and feelings that represents the inner plot of subjectivity—that is, the inner plot of what characterizes us as human beings.

Undoubtedly, Silvano Arieti's writings on cognition in the 1960s reveal an attempt to develop a highly formalized theoretical synthesis. The style of these writings is more difficult and less readable than his previous writings, in which Arieti tied his theoretical concepts to his clinical practice in clear, plain prose. His ideas on cognition, however, were not simply abstract theorizing; they were applicable to the therapeutic practice and to the very goal of psychotherapy—focused on the unfolding of the Ego and of conscious thought processes rather than on unconscious motivation. In this matter, Arieti disagreed with Paul Federn, who maintained that the treatment had to be aimed at "a repression of the Id rather than at the unfolding of the Ego" (SAP 1960a).

The most meaningful example of Arieti's cognitive approach in therapy is perhaps found in his treatment of schizophrenic hallucinations and delusions with a technique that consisted "chiefly of making the patient aware of certain processes that he himself brings about or over which he retains some control." Here, little by little, Arieti taught his patients to sequence the hallucinatory process into its different phases, to recognize the different stages of its unfolding, to identify the so called "listening attitude"—i.e., the phase when the patient is about to experience the hallucination—and to focus on it, in order to avoid its further development (Arieti 1962b, p. 52). In other words, the sequencing of thought to which Arieti referred in his writings about microgeny was put to good use in his psychotherapeutic practice.

In addition, Arieti believed that *explaining* to patients the formal mechanisms of their paleologic thinking was even more important than *interpreting* the contents of their hallucinations. As he said, "it should be explained to the patient how delusions and hallucinations are concretizations of what otherwise would be even more unbearable abstract concepts. For instance, a patient who had the olfactory hallucination that a rotten odor emanated from his body understood that the symptom was a concrete representation of the way he felt toward himself: he thought he has a smelly, rotten character" (Arieti 1960a, pp. 117–118). Returning to the theme of language, Arieti remarked that the therapist's task was similar to a translator's—helping a patient to become aware of how "he translates his psychodynamic conflicts into abnormal phenomena … Then, we help the patient to retranslate the symptoms in non-psychotic [ways]" (SAP 1965a).

This repeated emphasis on awareness was based on the belief that a patient's personality always retains an adult part, regardless of how disturbed he is. Arieti borrowed this idea from Frieda Fromm-Reichmann, who, in therapeutic settings, used to take on the role of "ambassador of reality"—a term coined by Paul-Claude Racamier, a French colleague (Racamier 1959). Arieti found this attitude inspirational, and it markedly distinguished his approach from that of other therapists like John Rosen and Marguerite Sechehaye, who, albeit in different ways, tried to reach their patients by entering into their psychotic world and aiming at a direct communication with their unconscious (Arieti 1962c; SAP 1971c). In this regard, Arieti stressed also his distance from existential psychiatric approaches. Taking into account the work of Ludwig Binswanger, in 1960 he wrote:

> In reading Binswanger's works, and those of other existentialists, I am reminded of works on esthetics, especially on poetry or on the theater. Now I myself like poetry and the theater very much, and furthermore I feel that we psychotherapists have unduly neglected the field of esthetics. But I also feel that esthetics alone is not enough. The person who studies a poem or a work of art tries to understand the work of art ... but he does not want to change it. One has the same feeling in reading these existentialist reports—the feeling that no attempt is made to change the patient. ... Everything is seen as fitting the world design, a superior symmetry. It is true that the existentialist tries to understand the uniqueness of the subjective experience of the patient, the so-called *Erlebnis*, but ... if we remain at this esthetic contemplation when we deal with psychiatric patients, we often really misplace our esthetic sense. ... Reading reports like the one I have summarized, one gets the feeling that the patient remains alone, in spite of the attempt made to understand the uniqueness of his experiences. As a matter of fact, we have the feeling that the existentialist psychiatrist does not participate at all. Like a spectator in the theater, he is not apathetic; he feels, admires, and suffers, but does not participate. He remains in the audience (Arieti 1960b, pp. 18–19).

Thus, Arieti's work was not headed towards an esthetic of illness; instead, it was headed towards an ethic of intervention that relied not only on relatedness but also on a modification of thought patterns (Arieti 1962c). This perspective was criticized as too normative by an anonymous reader who faulted Arieti's unwillingness "to investigate ... the different possible ways of experience, such as that of Heidegger's *Being-Here* [*Dasein*] as *Being-in-the-World* [*In-der-Welt-sein*], or such as that of Binswanger's *World-Project* [*Weltentwurf*]" (SAP Uh, my translation). Nevertheless, this feature of Arieti's approach showed his firm adhesion to medical practice, as well as his conviction about the power of the ideas to change one's life.

It is evident that during the 1960s Arieti was involved in a deep meditation about the future of psychotherapy and its renewal, a subject to which he paid much attention in his teaching at the William Alanson White Institute and at New York Medical College. The training of future therapists was very important to him, so much that in 1968 he suggested the institution of a special Board of Psychotherapy, to be attached to the American Board of Psychiatry and Neurology (Arieti 1968b). At this time he also began to treat depression, in which he adopted a dynamic-cognitive approach. His therapy of depression proceeded by sequencing the pathological stream of thoughts and guiding a patient "to recognize that the fit of depression comes as the result of the following ... sequence: *I am not getting what I should—I am deprived—I am in a miserable state*" (Arieti 1962d, p. 402). He guided patients to stop at the first stage of this sequence and to substitute for these recurring ideas different ones, reorganizing patients' ways of thinking so that "the usual clusters of thoughts" (Arieti

1962d, p. 403) would not recur and would not reproduce the old sequence. In short, Arieti again brought the concept of microgeny to his psychiatric practice, explicitly bringing it into play by choosing the same expression, "clusters of thoughts," that he had already used in his writings on microgeny. Once again, his theoretical conceptions were firmly linked to clinical practice.

4.3 Identity

In October 1968, at the third international Symposium on Feelings and Emotions, held in Chicago, Arieti participated with "Cognition and Feeling," a paper in which he reversed the traditional priority of these two elements. Whereas psychiatric literature generally affirmed the crucial role played by emotions, motivation, and feelings, he emphasized the dynamic role of ideas. In particular, he pointed out the confusion on this matter in psychiatry, a confusion that stemmed from the general assumption of "proto-emotions" ("first order emotions" that consist of an "inner status of the organism") as representative of the whole emotional domain. Psychiatry had neglected another class of emotions, those shaped and sustained by ideas, and these emotions concern the highest levels of the psyche and are connected with the progressive development of symbolic functions. In short, from the dynamic point of view, ideas have to be considered as more powerful forces for human beings than basic needs and feelings (SAP 1968h; Arieti 1970).

By stressing the close connection between thought and emotion and by reaffirming the role of cognition, Arieti was restating the importance of scientific reasoning in opposition to the irrational drift of the nineteenth and twentieth centuries—a move reminiscent of the call to rationality and to the "analysis of ideas" by past founders of human science, as, for example, by the eighteenth century French philosophers known as *Idéologues* (Gusdorf 1960; Moravia 1974; Kennedy 1978; Staum 2016).

Arieti's new book, *The Intrapsychic Self,* published in 1967, followed this line of argument (Arieti 1967). It is a difficult work, reflecting Arieti's belief that "there are rational forms of inquiry that are not those of experimental sciences" (SAP 1960d). Its framework differs greatly from that of *Interpretation of Schizophrenia,* as he is dealing with a formalized theoretical discourse based on a wide range of scientific literature.

The book was the result of a prolonged and thorough study, twelve years in the writing (SAP 1960e, 1968a). Aware of its complexity, Arieti decided to write a brief "guide to the book" for prospective readers:

> After the initial plans, I soon discovered that the work ahead was vaster than I had expected. Connections and implications sprang from all sides, and required digressions and the attainment of secondary goals. Not just two, three or four years passed by, ... but nine, ten, eleven. ... A fact became clear: that I had from time to time to revise ... the ground that I had originally intended to cover. ... In correlating a large amount of material I had to present to the reader well-known and familiar notions together with others, less known and at times difficult to assimilate. The reader ... will therefore encounter some unevenness, which my efforts could

> not smooth. To obviate these difficulties, I would like to suggest various ways in which the book can be read. The reader who is particularly interested in special topics ... may start by reading the chapter that covers that particular subject. Each chapter retains a partial autonomy which permits an adequate understanding without recourse to previous chapters. ... A second way consists of reading the book from the very beginning, but skipping all the material that does not arouse the interest of the reader. ... The third way of approaching the book, and the one that I wish will be adopted, consists of reading the whole volume from cover to cover. Then, the unevenness I referred to will be recognized as a secondary characteristic and, hopefully, the underlying unitary principles of the work will stand out (SAP 1966j).[5]

Soon after the publication of the book, Arieti sent a copy to an Italian friend and colleague, Mara Selvini Palazzoli, who was astonished at the amount of work he had undertaken. "I enviously wonder how the devil you do it," she wrote fondly; "I'll have to take a good deal of time to think about it. It seems to me extremely deep" (SAP 1967h, my translation). According to Selvini Palazzoli, the book had the merit of promoting theorization at a time when theory was increasingly neglected by psychiatry in favor of the promises of a golden age of psychopharmacology.

Placing theorizing about human nature at the core of psychiatry was actually the book's greatest challenge. In those pages Arieti described the intrapsychic self as "the core from which whatever is human expands and irradiates," and he connected its exploration with the search "of an understanding of what we are" (Arieti 1967, p. vii). Accordingly, Arieti wanted the title of the book to be *The Core of Man*, but eventually it was changed by the publisher (SAPd; SAPe).

"What is man?" (Arieti 1967, p. vii) What makes us who we are? In trying to answer these questions, Arieti assigned the primary role in human nature to cognition, since, according to him, cognition is what distinguishes us from other animals. Without ideas, man "would be limited to the dark core of I. He would be entirely a biological entity and not a sociobiological entity as we know him" (Arieti 1956, p. 37). Thus, thought is the distinguishing mark of human identity—an identity described as a "circular system" (Arieti 1967, p. 25) because it can transcend nature as a consequence of processes of thinking that are themselves rooted in our physiological nature.

Arieti's deliberation on cognition and thought merged into the question of the relationship between nature and culture, a subject with which he had already dealt in his early studies. To explain his point of view, Arieti began his book with a reference to Gordon Allport's distinction of two ways of conceiving human beings:

> As Allport wrote, there are two ways of studying man: one approach follows the Leibnizian tradition, the other the Lockean. The ... schools which follow the Leibnizian tradition see man predominantly as ... a psychological entity already equipped with full human status at birth. ... According to this point of view, man is endowed with biological equipment by which he makes contact with the world. ... In this approach the focus is on the intrapsychic ...
>
> The ... schools which follow the Lockean tradition see man's psyche basically as ... a *tabula rasa* at birth, an entity which is molded gradually by the experiences of life... Related to this approach is the conception that society and culture, rather than being made by man, are among the makers of man. ... In this tradition the focus is on the interpersonal.

[5] This text was only partially reproduced in the book's Preface.

Three fundamental positions taken by this author are that (1) man must be studied through both approaches, (2) some of the richest forms of human development are in the realm of the interpersonal, (3) the interpersonal presupposes an intrapsychic core (Arieti 1967, pp. 3–4).

The nature/culture question was therefore at the basis of a structural paradox of the human condition that clearly emerged from Arieti's circular reasoning. "The experience of the inner world is a vehicle for understanding the external world," he said; but, at the same time, "the experience of the external world is a vehicle for understanding the inner world" (Arieti 1967, p. 148). This circularity was noted in Benjamin Wolman's comment about the book. Upon reading a draft, Wolman admitted being impressed with Arieti's combination of Leibnizian and Lockean approaches. "The truth lies in a synthesis," Wolman summarized; "a man must be himself to be able to relate to others, but only in the relationship with others may he become himself" (SAP 1966c).

In earlier theoretical papers Arieti had suggested a conception of man as the result of a circular process of exchange between nature and culture (Arieti 1956, 1957). Now, more than ten years later, *The Intrapsychic Self* re-launched this conception. He was moving closer to a "unifying concept of man" by focusing on cognition as an integrative function of the elements of a multifaceted human nature. He therefore considered cognition as a third dimension, in addition to Leibnizian and Lockean dimensions that Allport had distinguished.

With its developmental approach, *The Intrapsychic Self* described the unfolding of cognition from its basic, most primitive, biological levels to the highest intellectual levels of creativity in art and science. With due respect to the Neo-Freudian school, in those pages Arieti connected his thinking to biology; according to him, it was not possible to envisage any development of intellectual processes without considering their origins in sensation, which had to be considered not only as the most elementary step of cognition but also as a biological function. Thus, cognition originates in biological processes. For example, Arieti described induction in the following way:

As conceived by Mill, induction is that type of inference by which we attempt to reach a conclusion concerning all the members of a class from observation of only some of them ... Only the organism that can perform inductively and transmit this inductive functioning genetically can survive. Not every association to which living forms were exposed became an unconscious induction in phylogeny, but only the small number with high survival value. This fact of selection shows that the validity of induction ... is in reference to survival and not to a philosophical absolute. It implies that some kind of regularity or uniformity of nature exists (Arieti 1967, pp. 136–137).

A few pages later, discussing deduction, he wrote:

We can repeat here what we have said about induction and state that a deductive procedure is followed unconsciously by the functioning organism. Only organisms which can operate in this way can survive. For instance, all bacteria coli produce an antibody reaction in the human organism. In a given case a particular strain of bacteria coli invades the organism: the organism will then produce antibody defenses and will survive. Immunity is based on the capacity of the organism to incorporate a deductive procedure. The difference between the organism and the psyche is that the psyche is able to subjectivize; that is, to become aware of the mechanism (Arieti 1967, p. 149).

Along with the biological origins of cognition, in those pages Arieti took into account the biological origins of human identity. Venturing into theoretical biology, he began with the elementary vital systems and speculated on the existence of an original, undifferentiated universe in which the first defined reactions of a primordial organism originated in its encounter with another organism—the two organisms different from one another but similar in that they possess a common vital nature.

According to Arieti, it was the ability of the primordial organism (PO1) to recognize another organism (PO2) as something similar but distinct, in a "simultaneous recognition of difference and sameness" (Arieti 1967, p. 178), that foreshadowed "the human ability to conceive unities" (Arieti 1967, p. 174) and led to the transformation of an undifferentiated world into a universe of "finite entities" (Arieti 1967, p. 176). In other words, Arieti suggested that PO1 recognized itself because of its encounter with another organism. This incipient "self-awareness" came about because of an individual's inherent endowment of biological universals—what in Kantian terms would be an a priori category of knowledge. In fact, it was the original capability of PO1 to recognize PO2 as something similar as well as different that allowed PO1 to recognize itself. Perhaps it was this nexus of ideas that Arieti was talking about when he said that his book was "meant to be a prerequisite to the study of the interpersonal" (Arieti 1967, p. 4).

The connection between identity and otherness, as well as the circular process of self-definition previously debated along the lines of Martin Buber's philosophy and Sullivan's dynamic thinking was now explored at the level of basic vital processes, a view that suggested a connection between a biological underlying substrate and the interpersonal dimension, for it was "within the individual that the long journey leading to the dialogue is initiated" (Arieti 1967, p. 3). In this way, Arieti linked Buber's philosophy of I-Thou with the concept of inwardness, since "the I, or self … is not just another" (SAP 1979).

> Inwardness makes us reach for ourselves, inside, opens to us our inner life. I enter into a special dialogue with a special person, me, I face myself, speak to myself, and read myself. ... I have a special encounter, "I-I"; a whole universe opens up to me, my own universe. But this inner universe consists of cognitive structures with cognitive content. If not for cognition, I could not have an inner life, or perhaps I could have only a very limited one. ... Many cognitive forms have a double entity; they consist of what seems a psychological bifurcation. One branch of this bifurcation is interpersonal, reaching the other with a word, an idea, a complicated relation. The other branch is intrapsychic, and makes it possible to retain such an idea, attitude, disposition, within ourselves. When I acquire a new cognitive form, let us say a new word or a new concept, not only it is my otherness which expands, but also my inwardness. Not only do I have a new way to reach others, or new understanding to give to others, but to myself (SAP 1979).

It was not by accident, Arieti explained, that the title's key-word, *Self*, referred to "the individual as he sees himself, or even better, the encounter of the individual with himself" (SAP Ub).

The recognition of similarities and differences that Arieti discussed in elementary vital systems was, he thought, also the basis of the highest symbolic functions. In mathematics, for example, when he examined the question of the biological and psychological origins of numbers, he suggested the same type of recognition in the relation between the concepts of one and two. Disagreeing with Friedrich Frege's anti-psychological assumption, Arieti wrote:

> If an individual responds mentally to unity B (let us say a bird) as he responded to A (another bird), and knows that B is not A, he has discovered number two. He recognizes that he reacts to B as he did to A, and yet that B is not A: they belong to the same class but are not the same object. At the same time that he identifies, he distinguishes. In other words, the concept of two implies the simultaneous recognition of difference and sameness (Arieti 1967, p. 178).

The capacity for the recognition of similarity thus went hand-in-hand with the possibility of *identifying* and *distinguishing*, which Arieti assumed to be the basic mechanisms of cognition.

On this point, Arieti's conception of cognition recalls the issues discussed by Rudolph Carnap in *The Logical Structure of the World*, where the German philosopher posits that the recognition of similarities and differences is at the basis of the structure of thinking (Carnap 1967).[6] Though Arieti did not refer to Carnap's book,[7] there is evidence of a certain influence of logical empiricism on his thinking. In *The Intrapsychic Self*, for example, he quotes Philippe Frank's *Philosophy of Science* (Frank 1957) and in some handwritten notes he explicitly refers to the Vienna Circle's 1929 *Manifesto* (SAPb).

It must be said that Arieti agreed with some relevant basic concepts of this school of thought—the need to establish a connection among different areas of knowledge in order to develop a unified science, the importance of avoiding philosophical and metaphysical absolutes (which emerges in Arieti's thinking on induction), and a call for a new scientific philosophy. But there were also strong grounds for Arieti to distance himself from the Vienna Circle. First, he could not agree with the school's physicalism, that is, the idea that all sciences must be translated into the physical properties of their subject matter. He also rejected the elimination of values and ethical issues from science. "Choice of values," he wrote in his notes on logical empiricism—almost a reminder on one of the main reasons of his disagreement with this school. Another annotation in the same set of notes can be read in a similar vein: "We must accept even what we do not understand" (SAPb, my translation)—pronouncement by which he emphasized his distance from the idea that all enigmas or empirically unverifiable phenomena had to be excluded from science.

In psychiatry, for example, though there were many crucial questions that had not been definitively answered, these questions remained within the perimeters of scientific inquiry. One such question was how awareness, the transformation of a neurological process into consciousness, comes about. Even recognizing that "the psyche brings to awareness what was already implied in the living matter" (Arieti 1967,

[6] The German edition of this book had appeared in 1928, under the title *Der Logische Aufbau del Welt* (Weltkreis Verlag, Berlin).

[7] Arieti did not know the German language, so he could not have read the original edition of Carnap's book. It is also unlikely that he read the English translation, since it appeared in the year *The Intrapsychic Self* was published. Nevertheless, it is possible that he read the Italian translation, published in 1961. In any case, no mention of the book appears among bibliographical references of Arieti's book.

p. 196), is not to know *how* it happens. Despite all efforts of research on this subject—for example, those of John C. Eccles (Eccles 1953)—the precise mechanisms of the transformation were still to be discovered. "Until biochemists or neurophysiologists are able to explain this aspect of life, we must resign ourselves to consider basic awareness an unanalyzable phenomenon" (Arieti 1967, p. 16), Arieti wrote.

Thus, far from being solved, the relationship of mind and body remained an unsolved enigma, and, at least since Descartes, a source of discomfort to philosophers and psychiatrists (Arieti 1968c). Among the different epistemological suggestions that could be of use in framing this problem—albeit not to solving it—Arieti referred to von Bertalanffy's idea of isomorphism:

> Von Bertalanffy also has made a profound study of the mind-body problem. He writes that he does not believe that he will arrive at final solutions dear to the heart of philosophers, but that he may make some progress relevant to psychological theory and psychiatric practice. He postulates an isomorphism between the constructs of psychology and neurophysiology: not a naive isomorphism which implies a photographic similarity between psychological and brain processes but a correspondence between psychological events and special codes and programs which are mediated by the nervous system. He believes that neurophysiology and psychology can be further unified. No attempt should be made to reduce the one to the other; rather, one should seek a generalized theory which could be applied to both (Arieti 1968c, p. 1631).

Therefore, it was not to logical empiricism but to the theoretical framework of General Systems Theory that Arieti turned for a synthesis. Above all, he endorsed von Bertalanffy's aim of creating synergy between science and humanism in the study of man and of avoiding reductionist conceptions of subjectivity, that is, of a realm where the verbs "identify" and "distinguish" assume a reflexive meaning—*identify oneself* and *distinguish oneself*.

The strong linkage between thought and subjectivity renders Arieti's book a contribution altogether different from the study of cognition in Ulric Neisser's 1967 *Cognitive Psychology* (Neisser 1967), for, as Arieti wrote, the emergence of subjectivity signified that "man can no longer be equated with a machine or a cybernetic system" (Arieti 1967, p. 28). Thus, with his book Arieti marked his distance from the computational approach that characterized the cognitive science of the mind.

In 1967, just before sending his final draft to the publisher, Arieti asked von Bertalanffy to read it. "You will recognize throughout the book your influence … In spite of many details, there is a constant search for unitary principles," wrote Arieti (SAP 1967f). In fact, the book mirrors the strong theoretical vein that characterized von Bertalanffy's scientific work, since it aimed at providing psychiatry with a clinical-based philosophy of human nature. In this effort, the study of schizophrenia had played a crucial role, since it had been through this illness that Arieti meditated on the paradox of being caught halfway between nature and culture, identity and otherness, similarity and difference. The analysis of schizophrenic thought was the keystone of Arieti's studies on cognition because—as von Domarus taught—schizophrenic thinking was characterized by the transformation of similarity into identity and therefore by the loss of the dialectics between similarity and difference that was a feature of healthy cognition and of sane development of self-identity. The schizophrenic lacked the synthesis of identity and otherness, similarity and differences that characterizes the healthy human being and finds its highest expression in the creative person—one who has the ability to grasp new similarities and to discover new connections. "It is on the varying responses to similarity that the ultimate rise or fall of man depends," wrote Arieti on the last page of his book; we can either confuse similarity with identity, as the schizophrenic does; or integrate it with difference "to break the secret of the universal night and make a piece of understanding a piece of ourselves" (Arieti 1967, p. 453).

Undoubtedly, as mentioned in a letter from his friend and colleague Morton Reiser, Arieti seemed to proceed through his book without catching his breath as he offered the reader dense, difficult, and profound insights. But let us pause for a moment to reflect that Arieti, who arrived in a foreign country in his early twenties as a Jewish refugee, is telling us that the core of man rests on our ability to recognize ourselves as both similar and different from what surrounds us. Perhaps, this is what Morton Reiser meant in the letter when he noticed Arieti's "fervor" and "involvement with the material" in the book (SAP 1969k). Here we see again a connection between science and life, as Arieti's interest in cognition seems to have been a feature of his maturation before it was a scientific concern, since it is owing to cognition that a man "can understand the world and see his place in it" (Arieti 1965, p. 29), as the author himself did in the United States.

In short, one cannot help thinking that Arieti's concern about human identity stemmed not exclusively from a strictly scientific interest, but also from a drive to understand himself. The questions he addresses to psychiatry are questions he keeps asking himself and we should not be surprised for that, since, as he himself pointed out, "in any conception of man it is humanly impossible to transcend the human character of the conception. Therefore, any such conception is subjective, reflexive, circular … There is no escape from what some philosophers call the anthropocentric predicament" (SAP Ug).

The Intrapsychic Self was published at the end of 1967. Because of the publisher's smart advertising, sales went well (SAPf). Its success, however, did not compare with that of *Interpretation of Schizophrenia* or with that of the *Handbook*, as its highly philosophical reasoning, conveyed sometimes with scientific rigor and sometimes with poetical expressiveness, made it challenging to read.

The book earned him recognition as a profound and cultured thinker—a point of reference for those who refuse to simplify the study of human beings (Arieti 1968c). Shortly afterwards he received a tempting invitation to participate in founding the Institute of Cognitive Psychiatry in Shreveport, Louisiana, and to assume the post of director (SAP 1969a). Though he was honored to receive the offer, he declined, for he was "not yet psychologically prepared to leave New York" (SAP 1969h), where he had finally established solid roots. For the same reason, in 1968 he had already declined an invitation to become Chairman of the Department of Psychiatry at the University of Michigan, in Ann Arbor (SAP 1968d).

Arieti, who had become Chairman of the Psychoanalytical Faculty of New York Medical College (SAP 1969n), was actually looking for a New York-based academic position. In early 1969, he asked Francis Braceland—then Editor of *The American*

Journal of Psychiatry—about the competition for the chairmanship of the Department of Psychiatry at New York University, for he had heard that Sam Wortis was about to retire. Arieti wrote:

> I feel that my contributions to the field of psychiatry should enable me to be considered for that position. My books are well known not only in the United States ... and the *Handbook* is adopted as a standard textbook in many schools. From time to time I have been offered or invited to consider chairmanships; however, they were always in the South, in obscure places or where I felt I would be isolated from cultural life.
>
> The position at N.Y.U. would appeal to me greatly, but the trouble is that nobody has offered it to me, and I don't have the least idea as to how to pursue the matter. Being aware of your knowledge of the psychiatric world as well as of your wisdom, I am hereby asking you for some illumination. First of all, do you believe it is appropriate for me to aspire to that position, or should I consider the idea presumptuous, unrealizable, and therefore to be discarded a priori? Second, if you feel that the idea is within the realm of possibility, can you advise me on how to proceed?
>
> I think I have the ability for training, research, therapy, and for didactical organization, but I have no one to propose my name (SAP 1969b).

Francis Braceland, not only an esteemed colleague but also a close friend, replied immediately, advising him warmly against such a move, arguing that the sheer amount of managerial chores of an administrative job would be harmful to him. "I have regarded you always as a scholar, a teacher, a writer, a man who needs time to think and to write. For you to try to get into that situation down there … would be a devastating thing." In addition, Braceland warned him that the Department staff was a close-knit group of which Arieti was not a member, and therefore it might prove very difficult for him to take it over: "The reason Sam Wortis could do it was he grew up in Neurology at Bellevue with Foster Kennedy. He had gone to New York University. He is a boy from the neighborhood" (SAP 1969c). With Arieti's identity as a scholar and his independence from "team mindset" as a worry of Braceland, his advice was that of a caring friend: "My strong advice to you would be not to think twice about it. A research professorship or an associate professorship as a clinician and a teacher, those things I can understand, but to head that department would be something else, and I would advise against it" (SAP 1969c). He stated that he would nevertheless support him, whatever he decided.

Arieti hesitated for a while. At first he thought he would try, and he would like at least to be considered for that position (SAP 1969d, 1969e, 1969g, 1969i, 1969l). But his timid attempts to contact Wortis ended in nothing, as he gave a "rather irritated" reply to Arieti's overture. "I think that this reception has given *le coupe de grace* to my ambition" (SAP 1969m), he wrote to Braceland, following a brief talk with Wortis.

Though the rejection was undoubtedly a blow, Arieti gained a deeper self-awareness from this episode. "I have done a lot of thinking, and more and more I agree with your original suggestion," he confessed to Braceland. "I believe that a research professorship rather than an administrative position would be appropriate for me" (SAP 1969o).

Loyal to his inner core, he remained a scholar (Fig. 4.1).

Fig. 4.1 "A scholar, a teacher, a writer, a man who needs time to think and to write." Silvano Arieti, early 1970s (SAP, Library of Congress, Manuscript Division) © Courtesy of James Arieti—All rights reserved

4.4 Homeland

After leaving Pisa in 1939, Silvano Arieti settled in the United States, which became his adoptive country. Unlike his brother Giulio, who returned to Italy, Silvano decided to stay where he had struggled and suffered and eventually found his way. If it is true, as Marguerite Yourcenar wrote, that "the true birthplace is that wherein for the first time one looks intelligently upon oneself" (Yourcenar 1959, p. 32), America was his true birth place—where a boy became a man and came to know himself.

Deciding to remain in America, however, did not mean abandoning his Italian attachments. Scientifically and personally, Silvano Arieti maintained his position between two countries, as revealed in his continuing relationships with his natal homeland, not the least of which were in the Italian images and cadences that he brough to his adoptive English (Clemmens, Spiegel, Bieber & Di Cori 1982). His familiarity with his native language is clear from his notes, where he occasionally switches between Italian and English. Consider, for example, the switch to English in this (untitled) poem, seemingly dedicated to his work as a therapist:

Con le ombre vive di tanti occhi

Che ci prospettano

Piani diversi di ascolto ...[8]

[8] "With the shadows of many eyes, which present to us different levels of listening".

Nothing extraordinary
It could be the daily work
Of any one of us (SAP, Ue).

In spite of difficulties, by dint of hard work English eventually became his second language, so fully that in speaking Italian he would occasionally use a few ungainly Americanisms.[9] On the other hand, his new language also retained some gray areas, as he never did assimilate some grammatical details: "One of these days, when you have time, you still have to explain to me when we have to use *that* instead of *which,*" he wrote in 1975 to the freelance editor of the second edition of *Interpretation of Schizophrenia* (SAP 1975). His vocabulary also showed a few errors that were Italianisms—like the use of "fest" instead of "feast" in the above-mentioned letter to von Bertalanffy (SAP 1968c).

Close relationships with people in Italy remained a constant in Arieti's private and professional life. I have already touched on his deep bonds with childhood friends, with whom he renewed correspondence after the war. Their mutual letters were dense, revealing the moods of young men overwhelmed by history. "Here we live with throbbing hearts" (SAP 1948, my translation), wrote one, describing the political events and the social collapse of a country exhausted by war. Arieti wrote to them about his life: his daily strains and consolations, the birth of his sons, and his encounter with psychoanalysis. He often sent packages of food and medication. They were reunited in 1950, when Silvano Arieti first returned to Italy after the war, bringing with him ballpoint pens—a new invention he symbolically gave to his former classmates with whom he had written so many essays during school days (SAPa).

Correspondence with his parents was extensive. Intermittent, brief wartime messages were replaced by frequent, long letters about daily life on opposite sides of the ocean, from the reopening of the Bemporads' shop in Pisa to Silvano's ending his addiction to cigarettes. Despite the distance, they spent holidays together by means of letters. Ines Arieti never failed to send little gifts to her loved ones: handmade woolen vests for the children and a kerchief for Jane. Her gifts for Silvano were addressed to her boy, who had left in 1939, and to the man who was making his mark in America: a box of chocolates and a necktie (SAPa).

Silvano Arieti became also a personage for his relatives in America, so much so Giampaolo and Yolanda, who lived in Miami, facetiously appointed him "honorary nanny" of their son Giacomo, who had moved to New York (SAP 1958b). Silvano was very close also to Enrico and Vana Bemporad and their sons, his cousins Jack and Jules—introducing the former to the study of philosophy (SAP 1981) and the latter to psychiatry.[10]

---

[9] A recurrent Americanism in Arieti's Italian papers is the incorrect use of the word *eventualmente* (from the English "eventually") in place of *alla fine* ("finally," "lastly")—as in Italian *eventualmente* actually means "possibly, maybe".

[10] Jack Bemporad became an internationally renowned Rabbi. It was at Arieti's home, where he often baby-sat for David, that he discovered philosophical readings on the bookshelf of his elder

The relationships in Italy were also important for Arieti in terms of science. The solid neurological education he had received at the University of Pisa remained an integral part of his scientific core. In 1947, believing that American psychiatry could benefit from an increased acquaintance with Italian neuropathological research, Arieti offered Bernard Alpers, the President of the American Neurological Association, to write abstracts of Italian scientific articles for *The Archives of Neurology and Psychiatry* (SAP 1947). Later, in the 1950s, he took charge of the section devoted to Italian scientific literature in *The American Journal of Psychotherapy* (SAP 1954, 1955f), for which, in 1955, he reviewed Lucio Bini and Tullio Bazzi's *Psicologia medica,* a detailed study of "the organic substratum of the psyche," a study he greatly appreciated, since it offered "to the American reader many valuable points of view … too often forgotten in American textbooks" (Arieti 1955a, p. 187).

After the war, Bazzi himself was the conduit through which Arieti became acquainted with the state of Italian psychiatry. They met in Rome in 1952, at the First International Congress of Neuropathology, where Arieti presented a paper on "The Pineal Gland in Old Age" (Arieti 1954; Fig. 4.2).

Fig. 4.2 Silvano Arieti with his wife Jane and his sons David and James. On the ocean liner going to Europe on the occasion of the first International Congress of Neuropathology, 1952 (Personal Collection of James Arieti) © Courtesy of James Arieti—All rights reserved

cousin, with whom he used to talk about philosophical matters during their weekly lunches in New York. Jules Bemporad, Jack's younger brother, was introduced to psychiatric studies by his elder cousin. Together they wrote *Severe and Mild Depression,* published in 1978.

From then on, they engaged in an intense scientific exchange. Arieti sent him his articles and informed him of his forthcoming book on schizophrenia; Bazzi, for his part, kept him posted about his *Trattato di psichiatria* (Bini & Bazzi 1954–1967) and suggested some scientific Italian contacts (SAP 1953).[11]

Promoting the circulation of Arieti's works in Italy was a challenge. Whereas American psychiatry was poorly versed in organic and neurological research, Italian psychiatry was nearly completely devoid of the dynamic and psychological culture presented in Arieti's work (Passione 2016, 2018). As Bazzi pointed out, two different scientific traditions were at stake: "I understand your surprise for not finding in our work the names of Karen Horney, Sullivan and others, who are essentially unknown here; and you will understand my surprise that no mention in your book can be found of Jaspers, Mayer-Gross, Berze, Schneider, and others, who made major contributions to the study of schizophrenia" (SAP 1955c, my translation), he wrote to his American colleague. Their perspectives on the etiology of schizophrenia were also quite different, for Arieti ascribed them to psychological origins, whereas Bazzi adhered instead to organic, biological causes.

In 1955, soon after the publication of *Interpretation of Schizophrenia* (Arieti 1955b) Arieti made inquiries about whether it could be translated. Through his correspondence with Bazzi, he knew that his book would be a departure for Italian psychiatry, and for this reason it would be difficult to find a medical publisher willing to invest in a translation. Without despairing, Arieti pursued his aim by turning to an old acquaintance, Igino Spadolini, his former professor of physiology in Pisa, to whom he wrote:

> Dear Professor Spadolini ... I think that in Italy a book like this is absolutely needed. I do not say so from an excess of presumption, but because I am aware of the state of psychiatry in Italy. As you certainly know, Italian psychiatry is still almost exclusively ruled by the neurological school ... and it has almost completely been deprived of all those psychological tendencies which have enhanced psychiatry in other countries. Although the book does not give up its neurological basis ..., it tries to integrate the psychological tendencies which are pretty unfamiliar to Italians ... It is undoubtedly influenced by the psychoanalytic school, but it is an American psychoanalysis, that is, pruned of Freudian excesses (SAP 1955a, my translation).

Spadolini promptly replied and introduced Arieti to Luigi Macrì, an Italian publisher specialized in the translation of foreign medical books. In July of 1955 an agreement between Basic Books and Macrì was drawn up (SAP 1955b, 1955e).[12]

Arieti's choice of translator was Gaetano Roi, based at the Psychiatric Hospital of Padua, then directed by Ferdinando Barison (Sbraccia 1996; Gozzetti & Cappellari 2002; Baccaro & Santi 2007; Babini 2009; Migliorini 2020).[13] Roi had just published

---

[11] Bazzi introduced Arieti to leading psychiatric and psychological journals, such as *Rassegna di psichiatria* and *Archivio di psicologia*, the latter directed by Agostino Gemelli.

[12] Basic Books granted Macrì the exclusive right of publication in Italian, and Macrì accepted the clause about publishing the book within twenty-four months. The agreement stated also that the translation had to be accurate and faithful and that no abridgment or change was permitted without the written consent of the author and of the American publisher.

[13] At that time the Psychiatric Hospital of Padua was a leading light of the renewal of Italian psychiatry. It had been directed since 1947 by Ferdinando Barison, who devoted a great deal of

the first Italian monograph on the psychotherapy of schizophrenia (Roi 1955; Arieti 1955c). He was not only "competent in the field of literature about schizophrenia" (SAP 1955d, my translation), but he was also, in a certain sense, the only choice available, given the concentration of Italian psychiatry on biological factors and its corollary neglect of psychology and psychotherapy.

Still, complications lay on the horizon. In October 1955, Macrì informed Arieti that because of the poor state of the Italian book market he could not pay for the translation. "You would not believe how I have to struggle to sell 1000 copies of a book here! We must cut our costs as much as possible, and the cost for the translation is not negligible" (SAP 1955g, my translation). Though Arieti decided to pay for the translation out of his pocket (SAP Uc), there was another complication—that Roi worked very slowly and with linguistic hesitation. Since he was not a trained translator, he tended to translate the more difficult passages too literally (SAPc). The problems, however, were not simply linguistic but also conceptual and cultural, since some words, like "borderline," "insight," and "ingrown family," required a theoretical understanding and a grounding in psychology not yet available in Italy. It is not surprising, therefore, that in 1957, when the publisher finally received the manuscript of the translation, many parts had to be completely re-done (Passione 2018).

By December 1960 the process had become so bogged down that the contract with Macrì was cancelled (SAP 1960g). Arieti greatly regretted this lost opportunity, all the more because in 1955, shortly after he had signed the contract with Macrì, he received an appealing proposal from Einaudi, a prestigious Italian publisher. The anthropologist Ernesto de Martino, author of *The World of Magic*,[14] acknowledged by many as the Italian Lévi-Strauss (de Martino 1972; Charuty 2009; Zinn 2015), and at that time director of the Einaudi's series "Ethnological, Psychological and Religious Studies," had learned about *Interpretation of Schizophrenia* through Emilio Servadio, an Italian pioneer of psychoanalytic studies (Errera 1990; Magherini 1994). Immediately, de Martino wrote to Arieti:

> Dear Colleague ..., I think that an Italian translation of your work would be worth doing, for many reasons: in general, in Italy biological interpretations of schizophrenia prevail, and in general our traditional psychiatry ... is not familiar with a psychodynamic point of view ... We are very backward also in the field of good translations of foreign books. Besides, in addition to being of interest to psychologists and psychiatrists, your work is a stimulating contribution also for ethnologists, historians of religions, and folklorists ... Therefore, I intend to recommend an Italian translation of your book to Einaudi; but before doing so I would like to know if my intention meets your approval (SAP 1955h, my translation).

---

study to schizophrenia. Barison was a representative of the phenomenological school, which in those years played a crucial role in counterbalancing the typical emphasis of Italian psychiatry on biological factors.

[14] The book was published in Italy in 1948 under the title *Il Mondo magico*.

Unfortunately, this letter arrived too late (SAP 1956a, 1956b, 1956e).[15] Had it arrived earlier, it is likely that an Italian translation of the book would have appeared in Italy in the 1950s—not in a psychiatric book series, but in an anthropological one. This, of course, did not occur, and the translation did not appear until 1963; in the interim, *Interpretation of Schizophrenia* had a limited circulation in Italy; it circulated only among those who could read English and were interested in its innovative approach (Mora 1955).

One of Arieti's Italian readers was Giovanni Enrico Morselli, author in 1930 of an original study on mental dissociation (Morselli 1930; Eddison 1930; Borgna 2002). Upon reading *Interpretation of Schizophrenia*, Morselli wrote to the American author to congratulate him and to point out an ideal connection: "In conceiving schizophrenia one can respect organicism while also considering a patient's psyche, a consideration that I myself consider as being fundamental" (SAP Ud, my translation).[16] In his letter, however, Morselli also pointed out that some things were slowly changing in Italy. In fact, he added, at the forthcoming International Congress, to be held in Zurich in 1957, Italian psychiatry would be represented by two open-minded figures, Vito Maria Buscaino and Mario Gozzano, who might be able to help disseminate Arieti's work in Italy (SAP Ud; SAP 1958a).

Despite Morselli's optimism, however, the moment had not yet arrived, as evidenced by the cancellation of the 1957 Symposium on Psychotherapy organized by Mario Gozzano, Gaetano Benedetti, and Agostino Gemelli. Because of a general skepticism by psychiatrists in Italy toward psychotherapy, the initiative had received very few submissions and the symposium could not go on (Molaro 2018).

In 1959 circumstances had hardly changed, so much so that, writing to Arieti soon after reading the *American Handbook of Psychiatry*, Isidoro Tolentino, an Italian psychiatrist, pointedly said: "I can't help thinking of the difference … between your way of conceiving psychiatry and our own, which is psychiatric in name or pretension only. Translating your work into Italian would be very useful" (SAP 1959b, my translation).

The year 1959 nevertheless proved to be pivotal. The first Chair of Psychiatry, as a position independent from Neurology, was established at the University of Milan. This was a significant change, for previously the teaching of psychiatry in Italian universities had been subsumed under neurology in the courses of Clinic of Nervous and Mental Diseases. Carlo Lorenzo Cazzullo, the holder of the newly "independent" Chair, had been trained under Armando Ferraro at the New York Psychiatric Institute. Soon after receiving this post, he invited Arieti to teach a seminar on schizophrenia (SAP 1960b, 1960c).

In those years, however, the renewal of Italian psychiatry took place mostly outside the university, and a crucial role was played by the "Milan Group"—a group of

---

[15] In his reply to de Martino, Arieti suggested the publication of an Italian edition of Clara Thompson's *Psychoanalysis: Evolution and Development*, but de Martino suggested that he send the proposal to Cesare Musatti, the director of the Einaudi series of books on psychology.

[16] This undated letter can actually be dated between 1955 and 1957, since Morselli referred both to *Interpretation of Schizophrenia*'s first American edition and to the forthcoming 1957 meeting in Zurich.

independent psychiatrists seeking new theoretical tools for their craft. Pier Francesco Galli, a pupil of Gaetano Benedetti, was a member of the Milan Group. In 1959, supported by Feltrinelli publishing firm, Galli launched the first Italian book series dedicated to psychoanalysis and dynamic psychiatry, introducing Italian readers to foreign psychiatrists (among whom were Harry Stack Sullivan and Frieda Fromm-Reichmann) (Babini 2009). The series, entitled *Biblioteca di psichiatria e psicologia clinica*, was a challenging undertaking, with Galli constantly seeking out new books. Having learned from George Mora—a mutual colleague—about Arieti's decision to cancel his contract with Macrì, Galli immediately offered to include *Interpretation of Schizophrenia* in the series (SAP 1959a, 1961a, 1961b, 1962a). This time everything went smoothly, and an Italian edition finally appeared in 1963 (Arieti 1963b). Alda Bencini Bariatti, who completed her postgraduate training in America, where she also attended the William Alanson White Institute, was the translator. She was well acquainted both with Arieti's language and scientific and cultural references.

In 1962 Silvano Arieti was invited to be an honored guest at the first training course organized by the Milan Group (SAP 1962a, 1962b; Arieti 1962e),[17] where he met, among others, Marco Bacciagaluppi, who in 1964 joined him for a semester in New York. Bacciagaluppi and his wife, Maria Mazza, would become the principal Italian translators of Arieti's writings (Bacciagaluppi 2018).

Arieti collaborated again on several occasions with members of the Milan Group, and his contributions were so appreciated that in 1964 the Group awarded him the Gold Medal for the Development of Psychotherapy (SAP 1964h). His visits to Milan became regular, and he was often asked to visit patients there.[18] To do so, he had to obtain an Italian license, since he had left Italy in 1939 without taking the State examination to practice medicine (HAUP 1966).[19]

Trips to Italy also facilitated more frequent family gatherings. Since his brother Giulio moved to Milan in 1960, where his parents often visited him, they could all meet there (SAPa). This renewed togetherness suddenly ended in 1963, when Ines Arieti died unexpectedly (SAP 1963a). This was a particularly sorrowful phase for Arieti, as, at the same time, he and his wife Jane, after twenty-two years of marriage, were obtaining a divorce. Divorce proceedings came to an end in 1965 (SAPi). Shortly afterwards, Silvano Arieti met Marianne Thompson Love. She was from Elmira, New York, and had lived in Philadelphia and Poughkeepsie. In 1965 she moved to New York after a long trip to Indonesia and the Philippines (SAP 1962c). An artist with three daughters, she was "a pleasant and smart woman, with a warm human touch and a deep interest in the things of this world" (SAP 1966i, my translation). Arieti fell in love with her, and they married shortly after the divorce (SAP 1965d; Fig. 4.3).[20]

---

[17] Other participants were Gaetano Benedetti, Pier Francesco Galli, Franco Fornari, Leonardo Ancona, Mara Selvini Palazzoli, Silvia Montefoschi, Ugo Marzuoli, Berta Neumann, Virginio Porta, Tommaso Senise, Enzo Spaltro, and Fabrizio Napolitani.

[18] Among Arieti's archival documents there is a small collection of papers with the label "Milan patients." However, no further information is provided.

[19] Arieti obtained a first temporary license in 1964 and a permanent one in 1966.

[20] Wedding year can be gathered from Silvano Arieti's will, dated October 25, 1965.

Fig. 4.3 Silvano Arieti with Marianne in Villasimius, Sardinia, early 1970s (SAP, Library of Congress, Manuscript Division) © Courtesy of James Arieti—

In short, within a brief period, many important things happened in his "not too static personal affairs" (SAP 1967e). So, after the mournful loss of his mother and the collapse of his marriage, a new phase of renewal began. He moved to an apartment at 103 East, 75th street, Manhattan, which also was his office (SAP 1966a, 1969r). In this new and more peaceful stage of his life, which allowed him to cope better with his many commitments, Silvano Arieti was happy to welcome with "much kindness and affection" (SAP 1968g, my translation) his father, Elio Arieti, in the autumn of 1968. Unfortunately, Elio, like his mother in 1963, unexpectedly died shortly afterwards, in January 1969, when he returned to Pisa (SAPh).

Given that his brother Giulio was back in the United States (SAP 1968e), this bereavement might have become a breach of Arieti's "last bond with Italy," as Alda Bencini Bariatti wrote in her affectionate letter of condolence (SAP 1969j, my translation). The situation turned out differently. After his father's death, Silvano Arieti decided to establish a stable base in Italy and bought a parcel of land in Villasimius, Sardinia; there he began construction of a house, which became his summer home (SAPl). Building the house was more than a symbolic endeavor, and more than symbolic was also that the cost of construction was paid in part with earnings from his collaboration with the *Enciclopedia italiana del Novecento,* to which the Italian physiologist Giuseppe Moruzzi invited him to contribute with an article on psychoanalysis (SAP 1969u; SAP 1970a, 1970f, 1971a, 1971b). Arieti's participation in this cultural enterprise of post-war Italy was a renewed and tangible bond with his homeland (Arieti 1980).[21] Another bond was made in 1969 with the publication of the Italian edition of the *American Handbook of Psychiatry* (Arieti 1969–1970), which was immediately successful (SAPj; SAP 1965b, 1965c, 1970b), and of the Italian edition of *The Intrapsychic Self* (Arieti 1969b).

The following year, he was invited to write an entry on schizophrenia for *The Encyclopedia Americana* (SAP 1970c, 1970d, 1970e; Arieti 1984), having already contributed an article on the topic in the *Encyclopaedia Britannica* (Arieti 1974).[22] His participation in works of such a major cultural and national value shows his solid roots on both sides of the Atlantic ocean. He was now a citizen of two worlds.

References

Allport, G. (1955). *Becoming: Basic Considerations for a Psychology of Personality.* Yale University Press.

Arieti, S. (1945). Primitive Habits and Perceptual Alterations in the Terminal Stage of Schizophrenia. *Archives of Neurology and Psychiatry,* 53 (5), 378–384.

Arieti, S. (1947). The Processes of Expectation and Anticipation. Their Genetic Development, Neural Basis and Role in Psychopathology. *The Journal of Nervous and Mental Disease,* 106 (4), 471–481.

Arieti, S. (1948). Special Logic of Schizophrenic and Other Types of Autistic Thought. *Psychiatry,* 11 (4), 325–338.

Arieti, S. (1954). The Pineal Gland in Old Age. *The Journal of Neuropathology and Experimental Neurology,* 13 (3), 482–491.

Arieti, S. (1955a). Book Review. Lucio Bini and Tullio Bazzi: Medical Psychology. Vallardi, Milano, 1954. *The American Journal of Psychotherapy,* 9 (1), 186–187.

Arieti, S. (1955b). *Interpretation of Schizophrenia.* Basic Books.

Arieti, S. (1955c). Book Review. The Psychotherapy of Schizophrenia. Gaetano Roi. *The American Journal of Psychotherapy,* 9 (4), 793–794.

Arieti, S. (1956). Some Basic Problems Common to Anthropology and Modern Psychiatry. *American Anthropologist,* 58 (1), 26–39.

Arieti, S. (1957). The Double Methodology in the Study of Personality and Its Disorders. *The American Journal of Psychotherapy,* 11 (3), 532–547.

Arieti, S. (Ed.). (1959–1966). *American Handbook of Psychiatry.* Basic Books.

Arieti, S. (1960a). Aspects of Psychoanalytically Oriented Treatment of Schizophrenia. In S.C. Scher & H.R. Davis (Eds.), *The Outpatient Treatment of Schizophrenia* (pp. 114–118). Grune & Stratton.

Arieti, S. (1960b). Recent Conception and Misconception of Schizophrenia. *The American Journal of Psychotherapy,* 14 (1), 3–29.

Arieti, S. (1962a). The Microgeny of Thought and Perception. A Psychiatric Contribution. *The Archives of General Psychiatry,* 6 (6), 454–468.

Arieti, S. (1962b). Hallucinations, Delusions and Ideas of Reference Treated With Psychotherapy. *The American Journal of Psychotherapy,* 16 (1), 52–60.

---

[21] The publication was delayed, and Arieti's contribution appeared in 1980 in the fifth volume of the Italian encyclopedia.

[22] This contribution appeared in the fifteenth edition of the Encyclopedia.

Arieti, S. (1962c). Psychotherapy of Schizophrenia. Some Theoretical and Practical Aspects. *The Archives* of *General Psychiatry,* 6 (2), 112–122.
Arieti, S. (1962d). The Psychotherapeutic Approach to Depression. *The American Journal of Psychotherapy,* 16 (3), 397–406.
Arieti, S. (1962e). Psicodinamica e psicoterapia delle psicosi schizofreniche. In Gruppo milanese per lo sviluppo della psicoterapia (Eds.), *Problemi di Psicoterapia 11–14 dicembre* (pp. 18–41). Centro Studi di psicoterapia clinica.
Arieti, S. (1963a). Studies on Thought Processes in Contemporary Psychiatry. *The American Journal of Psychiatry,* 120 (1), 58–64.
Arieti, S. (1963b). *Interpretazione della schizofrenia.* Feltrinelli.
Arieti, S. (1964). The Rise of Creativity: From Primary to Tertiary Process. *Contemporary Psychoanalysis,* 1(1), 51–69.
Arieti, S. (1965). Contributions to Cognition From Psychoanalytic Theory. In J. Masserman (Ed.), *Science and Psychoanalysis. Vol. III* (pp. 16–37). Grune & Stratton
Arieti, S. (Ed.) (1966). *American Handbook of Psychiatry. Vol. 3.* Basic Books
Arieti, S. (1967). *The Intrapsychic Self: Feeling, Cognition and Creativity in Health and Mental Illness.* Basic Books.
Arieti, S. (1968a). Some Memories and Personal Views. *Contemporary Psychoanalysis,* 5 (1), 85–89.
Arieti, S. (1968b). Further Training in Psychotherapy. *The American Journal of Psychiatry,* 125 (1), 96–97.
Arieti, S. (1968c). The Present Status of Psychiatric Theory. *The American Journal of Psychiatry,* 124 (12), 1630–1639.
Arieti, S. (1969a). Toward a Unifying Theory of Cognition. In W. Gray, F. J. Duhl, & N.D. Rizzo (Eds.) *General Systems Theory and Psychiatry* (pp. 193–208). Brown and Co.
Arieti, S. (1969b). *Il Sé intrapsichico.* Boringhieri.
Arieti, S. (Ed.) (1969–1970). *Manuale di psichiatria.* Boringhieri.
Arieti, S. (1970). Cognition and Feeling. In M.B. Arnold (Ed.), *Feelings and Emotions* (pp. 135–143). Academic Press.
Arieti, S. (1971). The Structural and Psychodynamic Role of Cognition on Human Psyche. In S. Arieti (Ed.), *The World Biennial of Psychiatry and Psychotherapy. Vol. 1* (pp. 3–33). Basic Books.
Arieti, S. (1972). *The Will To Be Human.* Quadrangle Books.
Arieti, S. (1974). Psychoses. In *Encyclopaedia Britannica* (pp. 173–179). Encyclopaedia Britannica, Inc.
Arieti, S. (Ed.) (1974–1975). *American Handbook of Psychiatry.* Basic Books.
Arieti, S. (1977). Cognitive Components in Human Conflict and Unconscious Motivation. *The Journal of the American Academy of Psychoanalysis,* 5 (1), 5–16.
Arieti, S. (1979). From Schizophrenia to Creativity. *The American Journal of Psychotherapy,* 33 (4), 490–505.
Arieti, S. (1980). Psicanalisi: il movimento psicanalitico. In *Enciclopedia del Novecento. Vol. V* (pp. 734–747). Istituto dell'Enciclopedia Italiana.
Arieti, S. (1984). Schizophrenia. In *The Encyclopedia Americana* (pp. 353–354). Grolier.
Arieti, S., Brodie, K. H. (Eds). (1981). *American Handbook of Psychiatry. VII. Advances and New Directions.* Basic Books.
Babini, V. P. (2009). *Liberi tutti. Manicomi e psichiatri in Italia: una storia del Novecento.* Il Mulino.
Baccaro, L., Santi, V. (Eds.) (2007). *Dai non luoghi all'esserci-con. Storie e testimonianze del Manicomio di Padova a cento anni dalla costruzione 1907–2007.* La Galiverna.
Bacciagaluppi, M. (2018). *Appunti autobiografici di uno psicoanalista relazionale.* Mimesis.
Barton, W. E. (1987). *The History and Influence of the American Psychiatric Association.* American Psychiatric Press, Inc.
Bemporad, J. (1981). In Memoriam. Silvano Arieti 1914–1981. *The Journal of the American Academy of Psychoanalysis,* 9 (4), iii–vii.
Bini, L., & Bazzi, T. (1954–1967). *Trattato di psichiatria.* Vallardi.

Borgna, E. (2002). Giovanni Enrico Morselli. In M. Maj & F.M. Ferro (Eds.), *Anthology of Italian Psychiatric Texts* (pp. 335–345). World Psychiatric Association.
Burhnam, D. L. (1976). Orthodoxy and Eclecticism in Psychoanalysis: the Washington-Baltimore Experience. In J.M. Quen, & E.T. Carlson (Eds.), *American Psychoanalysis: Origins and Developments* (pp. 87–108). Brunner/Mazel.
Carnap, R. (1967). *The Logical Structure of the World.* University of California Press.
Charuty, G. (2009). *Ernesto de Martino: Les vies antérieures d'un anthropologue.* Éditions Parenthèses.
Chomsky, N. (1957). *Syntactic Structures.* Mouton Publishers.
Clemmens, E., Spiegel, R., Bieber, I., & Di Cori, F. (1982). "Silvano Arieti: 1914–1981." *Academy Forum* 26: 6–9.
Conci, M. (2012). *Sullivan Revisited, Life and Work. Harry Stack Sullivan's Relevance for Contemporary Psychiatry, Psychotherapy and Psychoanalysis.* Tangram Edizioni Scientifiche.
Davidson, M. (1983). *Uncommon Sense: The Life and Thought of Ludwig von Bertalanffy.* J.P. Tarcher.
de Martino, E. (1972). *The World of Magic.* Pyramid Communications.
DeRosis, L. (1959). Arieti, Silvano, M.D., American Handbook of Psychiatry. *Library Journal,* 84, 3047–3049.
Eccles, J. C. (1953). *The Neurophysiological Basis of Mind.* Oxford University Press.
Eddison, H. (1930). G.E. Morselli, Mental Dissociation (*Sulla Dissociazione Mentale*). Rivista Sperimentale di Freniatria, 1930. *The Journal of Mental Science,* 76 (315), 840–841.
Errera, G. (1990). *Emilio Servadio. Dall'ipnosi alla psicoanalisi.* Nardini.
Flavell, J. (1963). *The Developmental Psychology of Jean Piaget.* Van Nostrand.
Frank, P. (1957). *Philosophy of Science.* Prentice Hall.
Gardner, H. (1985). *The Mind's New Science: A History of the Cognitive Revolution.* Basic Books.
Gozzetti, G., Cappellari, L. (2002). Ferdinando Barison (1906–1995). In M. Maj & F.M. Ferro (Eds.), *Anthology of Italian Psychiatric Texts* (pp. 347–363). World Psychiatric Association.
Grinker, R. (1969). An Essay on Schizophrenia and Science. *Archives of General Psychiatry,* 20 (1), 1–24.
Grinker, R. (1976). *Toward a Unified Theory of Human Behavior.* Basic Books.
Gusdorf, G. (1960). *Introduction aux Sciences Humaines.* Le Belles Lettres.
Hastings, D. W. (1960). American Handbook of Psychiatry, vol. 1 and vol. 2. Silvano Arieti ed. *Science,* 131 (3401), 656–657.
HAUP (1966). *Silvano Arieti. Personal File.* Tmin, October 26, (Estratto del Verbale della Sottocommissione per l'esercizio della professione di Medico Chirurgo).
JIMP (1960). The American Handbook of Psychiatry, edited by Silvano Arieti. *Journal of the Indian Medical profession* 6 (10): 2987.
Kennedy, E. (1978). *Destutt de Tracy and the Origins of "Ideology": a "Philosophe" in the Age of Revolution.* The American Philosophical Society.
Kilpatrick, M.E. (1961). American Handbook of Psychiatry. Volumes I and II. Edited by Silvano Arieti." *Mental Hygiene,* 45 (2), 292–293.
Laszlo, E. (1972). The Relevance of General Systems Theory. Papers Presented to Ludwig von Bertalanffy On His Seventieth Birthday. George Braziller.
Magherini, G. (1994). Servadio: un profilo intellettuale. *Inventario,* 3, 131–140.
Menninger, K. A. (1961). American Handbook of Psychiatry. 2 vols. Silvano Arieti. *Bullettin of the Menninger Clinic,* 25 (3), 266.
Menninger, K., Mayman, M. (1955). Episodic Dyscontrol: A Third Order of Stress Adaptation. *Bullettin of the Menninger Clinic,* 20 (4), 153–165.
Migliorini, G. (2020). Ferdinando Barison. In A. Molaro & G. Stanghellini (Eds.), *Storia della fenomenologia clinica. Le origini, gli sviluppi, la scuola italiana* (pp. 367–384). UTET.
Molaro, A. (2018). Ludwig Binswanger's Daseinsanalyse at the First Symposium on Clinical Psychology in Milan (1952). *European Yearbook of the History of Psychology,* 4, 37–102.

Mora, G. (1955). Recensioni: Arieti S. Interpretation of Schizophrenia. *Neuropsichiatria,* 11 (1), 128–129.
Moravia, S. (1974). *Il pensiero degli Idéologues. Scienza e filosofia in Francia 1870–1815*. La Nuova Italia.
Morselli, G. E. (1930). Sulla dissociazione mentale. *Rivista sperimentale di freniatria,* 54 (2), 209–322.
Neisser, U. (1967). *Cognitive Psychology.* Prentice Hall.
Passione, R. 2016. La psichiatria di Silvano Arieti: un primo profilo. *Physis. Rivista Internazionale di Storia della Scienza,* 51 (1–2), 219–330.
Passione, R. (2018). Language and Psychiatry: The Contribution of Silvano Arieti Between Biography and Cultural History. *European Yearbook of the History of Psychology*, 4, 11–36.
Pérez, R. M., Moskowitz, M., & Javier, R. (Eds.). (2004). *Reaching Across Boundaries of Culture and Class. Widening the Scope of Psychotherapy.* Rowman and Littlefield Publishers, Inc.
Racamier, P. C. (1959). Psychoanalytic Therapy of Psychoses. In S. Nacht (Ed.), *Psychoanalysis of Today* (pp. 171–204). Grune & Stratton.
Roi, G. (1955). *La psicoterapia nella schizofrenia.* Cappelli.
Sabshin, M. (1990). Turning Points in Twentieth-Century American Psychiatry. *The American Journal of Psychiatry,* 147 (10), 1267–1274.
Sabshin, M. (2008). *Changing American Psychiatry: A Personal Perspective.* American Psychiatric Publishing, Inc.
SAPa. *Correspondence, 1940–1981.*
SAPb. *Speeches and Writings, 1940–1981. Miscellany. Notes. Undated.*
SAPc. *Speeches and Writings, 1940–1981. Interpretation of Schizophrenia. First edition. Correspondence, 1954–1963.*
SAPd. *Speeches and Writings. The Intrapsychic Self: Feeling, Cognition and Creativity in Health and Mental Illness. Correspondence, 1967–1968.*
SAPe. *Speeches and Writings. The Intrapsychic Self: Feeling, Cognition and Creativity in Health and Mental Illness. Draft, circa 1966.*
SAPf. *Speeches and Writings. The Intrapsychic Self: Feeling, Cognition and Creativity in Health and Mental Illness. Promotion, 1967–1968.*
SAPg. *Subject File, 1914–1981. American Handbook of Psychiatry, editor. Correspondence.*
SAPh. *Subject File, 1914–1981. Personal. Arieti, Elio, 1914–1969.*
SAPi. *Subject File, 1914–1981. Personal. Divorce, 1964–1965.*
SAPj. *Subject File, 1914–1981. Publishing. Italian editors, 1957–1978.*
SAPk. *Subject File, 1914–1981. Ruby, Jack, 1963–1975.*
SAPl. *Subject File, 1914–1981. Villasimius property, 1969–1973.*
SAP (1947). L, November 28 (Bernard Alpers to Silvano Arieti). In *Correspondence, 1940–1981.*
SAP (1948). L, August 29 (Aldo Ballerini to Silvano Arieti). In *Correspondence, 1940–1981.*
SAP (1953). L, March 3 (Tullio Bazzi to Silvano Arieti). In *Correspondence, 1940–1981.*
SAP (1954). L, July 2 (Silvano Arieti to Emil Gutheil). In *Speeches and Writings, 1940–1981. Interpretation of Schizophrenia. First edition. Correspondence, 1954–1963.*
SAP (1955a). L, March 22 (Silvano Arieti to Igino Spadolini). In *Correspondence, 1940–1981.*
SAP (1955b). L, April 14 (Igino Spadolini to Silvano Arieti). In *Speeches and Writings, 1940–1981. Interpretation of Schizophrenia. First edition. Correspondence, 1954–1963.*
SAP (1955c). L, May 14 (Tullio Bazzi to Silvano Arieti). In *Speeches and Writings, 1940–1981. Interpretation of Schizophrenia. First edition. Correspondence, 1954–1963.*
SAP (1955d). L, May 14 (Silvano Arieti to Igino Spadolini). In *Speeches and Writings, 1940–1981. Interpretation of Schizophrenia. First edition. Correspondence, 1954–1963.*
SAP (1955e). T, July 14 (Agreement Between Basic Books and Macrì). In *Speeches and Writings, 1940–1981. Interpretation of Schizophrenia. First edition. Publishing, 1954–1955e.*
SAP (1955f). L, July 27 (Emil Gutheil to Silvano Arieti). In *Correspondence, 1940–1981.*
SAP (1955g). L, October 24 (Luigi Macrì to Silvano Arieti).

SAP (1955h). L, December 10 (Ernesto de Martino to Silvano Arieti). In *Speeches and Writings, 1940–1981. Interpretation of Schizophrenia. First edition. Correspondence, 1954–1963.*
SAP (1956a). L, January 21 (Silvano Arieti, Silvano to Ernesto de Martino). In *Speeches and Writings, 1940–1981. Interpretation of Schizophrenia. First edition. Correspondence, 1954–1963.*
SAP (1956b). L, January 30 (Ernesto de Martino to Silvano Arieti). In *Speeches and Writings, 1940–1981. Interpretation of Schizophrenia. First edition. Correspondence, 1954–1963.*
SAP (1956c). L, April 2 (Psychologist graduates to Institute Members). In *Correspondence, 1940–1981.*
SAP (1956d). L, June 20 (Silvano Arieti Silvano to Arthur Rosenthal). In *Subject File, 1914–1981. American Handbook of Psychiatry, editor. Correspondence.*
SAP (1956e). L, June 26 (Ernesto de Martino to Silvano Arieti). In *Speeches and Writings, 1940–1981. Interpretation of Schizophrenia. First edition. Correspondence, 1954–1963.*
SAP (1956f). L, July 12 (Silvano Arieti to Daniel Blain). In *Subject File, 1914–1981. American Handbook of Psychiatry, editor. Correspondence.*
SAP (1956g). L, July 12 (Silvano Arieti to Norman Cameron). In *Subject File, 1914–1981. American Handbook of Psychiatry, editor. Correspondence.*
SAP (1956h). L, July 21 (George Mora to Silvano Arieti).
SAP (1956i). L, September 4 (Silvano Arieti to Kurt Goldstein). In *Subject File, 1914–1981. American Handbook of Psychiatry, editor. Correspondence.*
SAP (1956j). L, September 21 (Silvano Arieti to Norman Cameron). In *Subject File, 1914–1981. American Handbook of Psychiatry, editor. Correspondence.*
SAP (1956k). L, September 26 (Arthur Rosenthal to Silvano Arieti). In *Subject File, 1914–1981. American Handbook of Psychiatry, editor. Correspondence.*
SAP (1956l). L, October 5 (Silvano Arieti to Frances Arkin).
SAP (1956m). L, October 5. (Silvano Arieti to David Engelhardt).
SAP (1956n). L, December 13 (Frances Arkin to Silvano Arieti).
SAP (1957a). L, January 29 (Harry Bone to Silvano Arieti). In *Subject File, 1914–1981. American Handbook of Psychiatry, editor. Correspondence.*
SAP (1957b). L, February 12 (Silvano Arieti to Daniel Blain). In *Subject File, 1914–1981. American Handbook of Psychiatry, editor. Correspondence.*
SAP (1957c). L, April 1 (Silvano Arieti to Norman Cameron). In *Subject File, 1914–1981. American Handbook of Psychiatry, editor. Correspondence.*
SAP (1957d). L, June 25. (Silvano Arieti to Franklin G. Ebaugh). In *Subject File, 1914–1981. American Handbook of Psychiatry, editor. Correspondence.*
SAP (1957e). L, October 22 (Silvano Arieti to Hervey Cleckley). In *Subject File, 1914–1981. American Handbook of Psychiatry, editor. Correspondence.*
SAP (1957f). L, December 10 (Silvano Arieti to Joseph B. Cramer). In *Subject File, 1914–1981. American Handbook of Psychiatry, editor. Correspondence.*
SAP (1958a). L, March 15 (Isidoro Tolentino to Silvano Arieti). In *Correspondence, 1940–1981.*
SAP (1958b). L, October 4 (Giampaolo Rocca to Silvano Arieti).
SAP (1959a). L, September 18 (Pier Francesco Galli to George Mora). In *Correspondence, 1940–1981.*
SAP (1959b). L, November 20 (Isidoro Tolentino to Silvano Arieti). In *Subject File, 1914–1981. American Handbook of Psychiatry, editor. Correspondence, 1956–1980.*
SAP (1960a). TMin, May 10 (Silvano Arieti, *Abstract of Round Table Meeting Held in Atlantic City. On Technical Aspects of Terminating Psychotherapy*). In *Speeches and Writings, 1940–1981. Lectures.*
SAP (1960b). L, June 16 (Carlo Lorenzo Cazzullo to Silvano Arieti). In *Correspondence, 1940–1981.*
SAP (1960c). L, June 27 (Silvano Arieti to Carlo Lorenzo Cazzullo). In *Correspondence, 1940–1981.*
SAP (1960d). L, September 21 (Bernard Kaplan to Silvano Arieti).

SAP (1960e). L, October 2 (Silvano Arieti to Arthur Rosenthal).
SAP (1960f). L, November 12 (Silvano Arieti to Arthur Rosenthal). In *Subject File, 1914–1981. American Handbook of Psychiatry, editor. Correspondence.*
SAP (1960g). L, December 19 (Arthur Rosenthal to Luigi Macrì). In *Speeches and Writings, 1940–1981. Interpretation of Schizophrenia. First edition. Correspondence, 1954–1963.*
SAP (1961a). L, January 21 (Silvano Arieti to Michele Ranchetti). In *Correspondence, 1940–1981.*
SAP (1961b). L, February 4 (Silvano Arieti to Pier Francesco Galli). In *Correspondence, 1940–1981.*
SAP (1961c). L, April 21 (Ralph Snyder to Silvano Arieti). In *Subject File, 1914–1981. Employment. Position appointments, 1941–1961c.*
SAP (1961d). L, May 16 (Silvano Arieti to Charles Kaufman). In *Correspondence, 1940–1981.*
SAP (1961e). L, May 22 (Charles Kaufmann to Silvano Arieti). In *Correspondence, 1940–1981.*
SAP (1961f). L, July 26 (Dale Cameron to Silvano Arieti).
SAP (1962a). L, June 3 (Pier Francesco Galli to Silvano Arieti). In *Correspondence, 1940–1981.*
SAP (1962b). L, November 19 (Alda Bencini Bariatti to Silvano Arieti). In *Correspondence, 1940–1981.*
SAP (1962c). M (Marianne Arieti, *Travel Journal*).
SAP (1963a). C, February 23 (Death certificate of Ines Bemporad). In *Subject File, 1914–1981. Personal. Certificates, 1940–1975.*
SAP (1963b). T, July 3, (Agreement By and Between Silvano Arieti and Basic Books). In *Subject File, 1914–1981. American Handbook of Psychiatry, editor. First edition. Vol. 3, 1963b–1966.*
SAP (1963c). L, August 28 (Silvano Arieti to Carl Rogers). In *Subject File, 1914–1981. American Handbook of Psychiatry, editor. Correspondence.*
SAP (1963d). L, September 4 (Ludwig von Bertalanffy to Silvano Arieti). In *Subject File, 1914–1981. American Handbook of Psychiatry, editor. Correspondence, 1956–1980.*
SAP (1963e). T, November 25 (John Holbrook, *Mental Status of Rubenstein, Jack, Alias Ruby, Jack*). In *Subject File, 1914–1981. Ruby, Jack, 1963e–1975.*
SAP (1963f). TL, December 18 (Henry Wade and William Alexander to Silvano Arieti). In *Subject File, 1914–1981. Ruby, Jack, 1963f–1975.*
SAP (1964a). L, January 16 (John Holbrook to Silvano Arieti). In *Subject File, 1914–1981. Ruby, Jack, 1963–1975.*
SAP (1964b). L, March 23 (John Holbrook to Silvano Arieti). In *Subject File, 1914–1981. Ruby, Jack, 1963–1975.*
SAP (1964c). L, April 13 (Silvano Arieti to John Holbrook). In *Subject File, 1914–1981. Ruby, Jack, 1963–1975.*
SAP (1964d). T, May 27 (Silvano Arieti, *The Rise of Creativity: From Psychopathology to Innovation*). In *Speeches and Writings, 1940–1981. Lectures.*
SAP (1964e). L, September 10 (Silvano Arieti to Alfred Freedman). In *Correspondence, 1940–1981.*
SAP (1964f). L, October 22 (Leon Salzman to Silvano Arieti). In *Correspondence, 1940–1981.*
SAP (1964g). L, November 4 (Silvano Arieti to Leon Salzman). In *Correspondence, 1940–1981.*
SAP (1964h). L, November 9 (Pier Francesco Galli to Silvano Arieti). In *Correspondence, 1940–1981.*
SAP (1965a). T, May 4 (Silvano Arieti, *The Psychoanalytic Approach to the Psychoses. Round Table*). In *Speeches and Writings, 1940–1981. Lectures.*
SAP (1965b). L, October 5 (Laura Schwarz to Silvano Arieti). In *Correspondence, 1940–1981.*
SAP (1965c). L, October 10 (Silvano Arieti to Laura Schwarz). In *Correspondence, 1940–1981.*
SAP (1965d). T, October 25 (Last Will and Testament of Silvano Arieti). In *Subject File, 1914–1981. Finances. Wills, 1963–1979.*
SAP (1965e). L, October 29 (Robert Robinson to Silvano Arieti).
SAP (1965f). L, November 16 (Silvano Arieti to Robert Robinson).
SAP (1966a). L, March 12 (Silvano Arieti to Rose Spiegel). In *Correspondence, 1940–1981.*
SAP (1966b). T, May 10 (Silvano Arieti, *Comments of Silvano Arieti, M.D., Presented at the 122nd Annual Meeting of the American Psychiatric Association. Program on Psychiatry and General System*). In *Speeches And Writings, 1940–1981. Reviews, 1956–1977.*

SAP (1966c). L, July 2 (Benjamin Wolman to Silvano Arieti). In *Speeches and Writings. The Intrapsychic Self: Feeling, Cognition and Creativity in Health and Mental Illness. Correspondence, 1967–1968.*

SAP (1966d). L, September 27 (Silvano Arieti to William Gray).

SAP (1966e). L, October 6 (Donald Oken to Silvano Arieti). In *Correspondence, 1940–1981.*

SAP (1966f). L, October 7 (Silvano Arieti to Donald Oken). In *Correspondence, 1940–1981.*

SAP (1966g). L, November 7 (Stanley F. Yolles to Silvano Arieti). In *Correspondence, 1940–1981.*

SAP (1966h). L, November 22 (William Gray to Silvano Arieti).

SAP (1966i). L, December 20 (Aldo Ghezzani to Silvano Arieti). In *Correspondence, 1940–1981.*

SAP (1966j). T (Silvano Arieti, *Preface and Guide to the Book*). In *Speeches and Writings. The Intrapsychic Self: Feeling, Cognition and Creativity in Health and Mental Illness. Draft, circa 1966j.*

SAP (1967a). L, March 14 (Francis Braceland to Silvano Arieti).

SAP (1967b). L, March 28 (Silvano Arieti to Francis Braceland).

SAP (1967c). L, April 19 (William Gray to Silvano Arieti). In *Correspondence, 1940–1981.*

SAP (1967d). L, May 4 (Silvano Arieti to William Gray). In *Correspondence, 1940–1981.*

SAP (1967e). L, May 8 (Silvano Arieti to Bernard Kaplan). In *Subject File, 1914–1981. American Handbook of Psychiatry, editor. Correspondence, 1956–1980.*

SAP (1967f). L, May 14 (Silvano Arieti to Ludwig von Bertalanffy). In *Speeches and Writings. The Intrapsychic Self: Feeling, Cognition and Creativity in Health and Mental Illness. Correspondence, 1967f–1968.*

SAP (1967g). L, November 24 (Silvano Arieti to John Schimel). In *Correspondence, 1940–1981.*

SAP (1967h). L, December 27 (Mara Selvini Palazzoli to Silvano Arieti). In *Speeches and Writings. The Intrapsychic Self: Feeling, Cognition and Creativity in Health and Mental Illness. Correspondence, 1967h–1968.*

SAP (1968a). L, January 14 (Silvano Arieti to Arthur Rosenthal). In *Speeches and Writings. The Intrapsychic Self: Feeling, Cognition and Creativity in Health and Mental Illness. Correspondence, 1967–1968a.*

SAP (1968b). L, February 7 (Silvano Arieti to David Rothstein).

SAP (1968c). L, March 18 (Silvano Arieti to Ludwig von Bertalanffy).

SAP (1968d). L, May 7 (Silvano Arieti to Thomas N. Cross).

SAP (1968e). L, May 7 (Elio Arieti to Silvano Arieti). In *Correspondence, 1940–1981.*

SAP (1968f). T, May 11 (Silvano Arieti, *Response to the Presentation of the Frieda Fromm-Reichmann Award*). In *Speeches and Writings, 1940–1981. Lectures.*

SAP (1968g). L, October 10 (Elio Arieti to Silvano Arieti). In *Correspondence, 1940–1981.*

SAP (1968h). T (Silvano Arieti, *Cognition and Feeling*). In *Speeches and Writings, 1940–1981. Lectures.*

SAP (1969a). L, January 16 (Walter Bruschi to Silvano Arieti). In *Correspondence, 1940–1981.*

SAP (1969b). L, February 21 (Silvano Arieti to Francis Braceland).

SAP (1969c). L, February 25 (Francis Braceland to Silvano Arieti).

SAP (1969d). L, March 7 (Silvano Arieti to Francis Braceland).

SAP (1969e). L, March 12 (Francis Braceland to Silvano Arieti).

SAP (1969f). L, March 13 (Harold Lief to Silvano Arieti).

SAP (1969g). L, March 17 (Silvano Arieti to Francis Braceland).

SAP (1969h). L, March 18 (Silvano Arieti to Walter Bruschi). In *Correspondence, 1940–1981.*

SAP (1969i). L, March 19 (Francis Braceland to Silvano Arieti).

SAP (1969j). L, March 26 (Alda Bencini Bariatti to Silvano Arieti).

SAP (1969k). L, March 27 (Morton Reiser to Silvano Arieti). In *Correspondence, 1940–1981.*

SAP (1969l). L, April 1 (Francis Braceland to Silvano Arieti).

SAP (1969m). L, April 11 (Silvano Arieti to Francis Braceland).

SAP (1969n). L, April 14 (Silvano Arieti to Alfred Freedman).

SAP (1969o). L, June 2 (Silvano Arieti to Francis Braceland).

SAP (1969p). L, June 10 (Frederik Duhl to Silvano Arieti).

SAP (1969q). L, August 12 (William Gray to Silvano Arieti). In *Correspondence, 1940–1981.*
SAP (1969r). L, October 24 (Marianne Arieti to W.E. Crichton). In *Correspondence, 1940–1981.*
SAP (1969t). L, December 13 (Silvano Arieti to Arthur Rosenthal). In *Subject File, 1914–1981. American Handbook of Psychiatry, editor. Correspondence.*
SAP (1969u). L, December 17 (Silvano Arieti to Giorgio Macchi).
SAP (1970a). L, January 16 (Giuseppe Moruzzi to Silvano Arieti).
SAP (1970b). L, February 24 (Leonardo Ancona to Mara Selvini Palazzoli). In *Correspondence, 1940–1981.*
SAP (1970c). L, June 16 (Loring Batten to Silvano Arieti). In *Correspondence, 1940–1981.*
SAP (1970d). L, June 18 (Silvano Arieti to Loring Batten). In *Correspondence, 1940–1981.*
SAP (1970e). L, July 13 (Silvano Arieti to Loring Batten). In *Correspondence, 1940–1981.*
SAP (1970f). L, December 8 (Silvano Arieti to Giorgio Macchi).
SAP (1971a). L, January 20 (Silvano Arieti to Vincenzo Cappelletti).
SAP (1971b). L, February 27 (Silvano Arieti to Vincenzo Cappelletti).
SAP (1971c). T (Silvano Arieti, *Revisiting the Concept of Transference and Counter-Trasference in the Psychoanalytic Therapy of Schizophrenia*). In *Speeches and Writings, 1940–1981. Lectures.*
SAP (1974a). L, October 22 (William Gray to Silvano Arieti). In *Correspondence, 1940–1981.*
SAP (1974b). L, October 29 (Silvano Arieti to William Gray). In *Correspondence, 1940–1981.*
SAP (1975). L, May 13 (Silvano Arieti to Raymond Mullaney). In *Speeches and Writings, 1940–1981. Interpretation of Schizophrenia. National Book Award, 1975.*
SAP (1979). T, May 11 (Silvano Arieti, *Cognition in Psychoanalysis*). In *Speeches and Writings, 1940–1981. Lectures.*
SAP (1981). L, March 23 (Jack Bemporad to Silvano Arieti). In *Correspondence, 1940–1981.*
SAP (Ua). HN, Unt. (Notes on Jack Ruby). In *Subject File, 1914–1981. Ruby, Jack, 1963–1975.*
SAP (Ub). HN, Unt. (Notes on the concept of Self). In *Speeches and Writings, 1940–1981. The Intrapsychic Self: Feeling, Cognition and Creativity in Health and Mental Illness. Notes. Circa 1966.*
SAP (Uc). L (Silvano Arieti to Luigi Macrì).
SAP(Ud). L (Giovanni Enrico Morselli to Silvano Arieti). In *Correspondence, 1940–1981.*
SAP (Ue). P, Unt. In *Subject File, 1914–1981. Miscellany, 1962–1981.*
SAP (Uf). T, Unt. (Notes for a Speech on Pluralism in Psychiatry).
SAP (Ug). T (Silvano Arieti, *The Conception of Man*). In *Speeches and Writings, 1940–1981. Articles. Undated.*
SAP (Uh). T (UA, *Problemi di traduzione e interpretazione dell'esperienza pre-logica*).
Satten, J., Menninger, K., Rosen, I., & Mayman, M. (1960). Murder Without Apparent Motive: A Study in Personality Disorganization. *The American Journal of Psychiatry,* 117 (1), 48–53.
Sbraccia, F. (Ed.) (1996). *Schizofrenia: labirinti e tracce.* La Garangola.
Shaw, M. (2007). *Melvin Belli: King of the Courtroom.* Barricade Books.
Staum, M. (2016). *Cabanis. Enlightenment and Medical Philosophy in the French Revolution.* Princeton University Press.
Steinberg, D. (1960, November 9). Basic Books Venture Becomes Big Business. *The New York Herald Tribune,* 41.
Taschdjian, E. (Ed.). (1975). *Perspectives on General Systems Theory. Scientific-Philosophical Studies.* George Braziller.
Ullman, M. (1965). Cognitive Processes and Psychopathology. *Science,* 147 (3660), 914–915.
von Bertalanffy, L. (1964). The Mind-Body Problem: A New View. *Psychosomatic Medicine,* 26 (1), 29–45.
von Bertalanffy, L. (1965). *General Theory of Systems. Application to Psychology.* Unesco Digital Library. https://unesdoc.unesco.org/ark:/48223/pf0000156058?2=null&queryId=75165110-10b7-44e4-b670-5de7509d3aba
von Bertalanffy, L. (1968a). *General System Theory: Foundations, Development, Applications.* George Braziller.

von Bertalanffy, L. (1968b). *The Organismic Psychology and Systems Theory. Heinz Werner Lectures.* Clark University Press.
von Bertalanffy, L. (1969). *Robots, Men and Minds: Psychology in the Modern World.* George Braziller.
von Bertalanffy, L. (1981). *A Systems View of Man: Collected Essays.* Westview Press.
Yourcenar, M. (1959). *Memoirs of Hadrian.* Penguin Books.
Zinn, D. L. (2015). An Introduction to Ernesto de Martino's Relevance for the Study of Folklore. *The Journal of American Folklore,* 128 (507), 3–17.

Chapter 5
Horizons

5.1 Expanding the Boundaries

Despite his father's death—or perhaps *because of* it—in the 1970s Silvano Arieti strengthened his bonds with Italy. Frequent correspondence with the friends of his youth continued throughout the decade (SAPa)—a connection that mirrors in Arieti's private life the importance of the interpersonal perspective that was the basis of his psychiatric thinking. Though the years were passing, he seemed to gain a sense of cohesiveness with his correspondents. One wrote, "You have the same handwriting that you had forty years ago. I think this suggests that you maintain the same balance as then. You are likely still the incurable optimist you have always been; truly, you have preserved your faith in mankind" (SAP 1974c, my translation).

His correspondence with the affectionate Tina, who had been housekeeper at the Arieti house in Pisa until Elio's death, was also intense. Her frequent letters to Silvano conveyed a sense of gloomy nostalgia. "When I go to Pisa I am troubled, for I always think of the good days spent with your parents when you and Giulio were kids," she wrote. Then she continued: "Yet, is this what life is like? Dying or seeing others die?" (SAP 1976b, my translation).

The answer was inescapable: seeing others die and continuing to live, as Arieti did, projecting himself ceaselessly, almost feverishly, into the future, as though not having enough time. "He was menaced by the flight of time; he could always hear the clock ticking, he could never fully relax," his son James remembers. Even at the restaurant, "if the waiter did not bring the menu immediately, he would grow antsy" (Arieti 2001, p. 35). Nevertheless, he took his time to correspond with his friends and relatives (SAPa).

During the 1970s and until his death in 1981, Arieti's annual summer visits to Sardinia were an opportunity to engage in scientific conversations with Italian psychiatrists. During his vacation (generally, from late July through most of August) he often lectured and taught seminars in Italian universities, where he was sought-after and valued, owing to the success of his *Manuale di psichiatria* (Arieti 1969–1970). At the University of Cagliari, he was a regular guest at the Psychiatric Institute directed

R. Passione, *Psychiatry and the Human Condition*, Springer Biographies,
https://doi.org/10.1007/978-3-031-09304-3_5

by Alfonso Mangoni, who would have liked to have him teach in Italy on a permanent basis (SAP 1972g). Another quite regular appointment was at the University of Naples, where Dargut Kemali and Raffaello Vizioli invited him to conduct seminars on psychodynamics (SAP 1973c). His lectures, dedicated to the psychopathological aspects of psychoses, were the only ones that employed the psychodynamic approach instead of an organic one (SAP 1976a, 1976e, 1976f).

Arieti's perspective was also introduced at the University of Bologna, where he was invited by Carlo Gentili, who wanted his students to become familiar with an approach capable of meeting the needs of the new community mental health care to which Italy was moving after the 1968 passage of the "Mariotti law" (Babini 2009, 2014; De Risio 2019). The law opened Italian psychiatry to a rethinking along theoretical and practical lines for which Arieti could make useful suggestions.[1] In particular, Gentili believed that the current transformation of Italian mental health care "would require a more refined knowledge of the possibilities of psycho- and socio-therapeutic intervention, a clearer awareness of achievable goals in out-patient care, and a prognostic evaluation of feasible therapeutic strategies" (SAP 1977c, my translation) in the field of community psychiatry.

Additionally, Arieti received invitations from university departments, mental health centers, and psychiatric institutions from all over the country. In 1972 he flirted with the idea of buying a house in Rome and opening an office with Gaetano Benedetti, but he did not follow through with this project (SAP 1972m).[2] Thus, well beyond the members of the "Milan Group"—with whom he continued to maintain professional relations (SAPa)—Silvano Arieti was acknowledged an important scientific figure in Italy during the 1970s, so much so that in 1977 he was invited by Arnaldo Forlani, the Italian Minister for Foreign Affairs, to a reception of the General Assembly of the United Nations (SAP Ua).[3]

His native homeland was not the only country with which Arieti established solid connections. From the end of the 1960s on he frequently traveled abroad. In 1969 he went to Buenos Aires, Argentina, to teach a series of classes to psychiatrists (SAP 1969a). In the same year he was a guest at the International Colloquium on Psychoses in Montreal, organized by the Institute Albert Prévost (SAP 1969b); he returned to Canada again in 1977 (SAP 1977d). In 1970 he traveled to England, where over a period of ten days he attended one conference after another (SAP 1970c). He then went to Zurich, where he participated in the Symposium on Schizophrenia presided by Manfred Bleuler, held on the occasion of the centenary of the Burghölzli Clinic (SAP 1970a). From there, he flew to Israel, where he had been in 1960 (SAP 1960,

[1] In 1968, promulgated by the Minister of Health Luigi Mariotti, Law number 431 opened up changes in Italy in psychiatric assistance – the idea of community psychiatry being acknowledged there for the first time.

[2] This project, about which Marco Bacciagaluppi informed me in one of our conversations, is confirmed in Arieti's correspondence with the "Giorgi Giotto" Real Estate.

[3] The invitation card reads "The Honorable Arnaldo Forlani, Minister of Foreign Affairs of Italy, Chairman of the Italian Delegation to the Thirty-second Session of the General Assembly of the United Nations, requests the pleasure of the company of Dr. and Mrs Silvano Arieti at a reception on Friday, 30 September 1977, at 6–7.30 o'clock".

1969d). In early August 1971 he was in Finland to give a speech in Turku at the Fourth International Symposium on the Psychotherapy of Schizophrenia (SAP 1971e). Here he ran into a mishap, as he suddenly developed acute neurological symptoms at first attributed to a cervical hernia but later diagnosed as shingles. He was hospitalized; at the end of August he was still "very weak and in great pain" (SAP 1971f) and seemed to require rest. Still, the shingles did not stop him. By September he felt well enough to board a plane for Mexico, where he took part in the second "Psychiatric Week" organized by Alonso Cantu Cantu, the President of the Sociedad de Psiquiatria of Monterrey (SAP 1971g).

His packed schedule kept him in motion for the full decade, with visits to Venezuela, Japan, and Norway (SAPa; SAPi). Everywhere he went he found new colleagues and met students who would then keep in touch with him, often trying to work with him in the United States (SAPa).

Arieti's international commitments were aimed at breathing new life into American psychiatry and increasing the eclecticism that had always been a distinguishing mark of his thought. In this effort he worked to establish solid channels between countries. In 1969, for example, with a view to future trips abroad, he deliberated with Gerard Chrzanowski and Lothar Kalinowski about founding a Center of Transcultural Psychiatry at New York Medical College. They spoke about it with the Chairman of the Department, Al Freedman, but apparently it did not meet his approval, and their plan went no further (SAP 1969c).

Another venture was more successful. It concerned a book—a collective work similar to the *American Handbook* but with an international character. "After this, you'll probably not want to edit another book again," a colleague had written in 1966, during the preparation of the Handbook (SAP 1966a). But he did: "I am always cooking something" (SAP 1968d), Arieti wrote to Charles Wahl, of the Department of Psychiatry of UCLA Medical Center, telling him about his new international project, the *World Biennial of Psychiatry and Psychotherapy.*

Silvano Arieti had been working on it since 1965, when he submitted the proposal for the project to Arthur Rosenthal of Basic Books. It was to be a biennial volume aimed at the widest possible dissemination of the most significant psychiatric developments from all over the world. The Editorial Board was to include American, Swiss, French, English, Italian, German, Spanish, Israeli, and Polish psychiatrists. "I am still working on the Canadians, South Americans, Russians, and Japanese" (SAP 1965), he observed.

Basic Books approved the project, and the contract for this "exciting new venture" (SAP 1966c) was drawn up in December 1966. Arieti immediately attended to the choice of members of the Editorial Board and of potential contributors. Some declined: Konrad Lorenz, who had already many commitments and was grappling with health issues (SAP 1966b, 1966d); Giuseppe Moruzzi, who was working on a chapter for Landois-Rosemann's *Lehrbuch der Physiologie des Menschen*, a time-consuming commitment (SAP 1967a); and William Masters, who, along with Virginia Johnson, had just published *Human Sexual Response* (Masters & Johnson 1966), because their University's Board of Directors prohibited them to publish outside the Reproductive Biology Research Foundation (SAP 1967b). Also Ronald Laing, who had accepted Arieti's invitation, later backed out (SAP 1968a).

The Editorial Board ultimately included leading psychiatrists from thirteen countries: Brazil, France, Germany, Israel, Italy, Japan, Mexico, Peru, Poland, Spain, Switzerland, Soviet Union, and United States.[4] In addition, the book included contributions from Irish, Norwegian, Canadian, Australian, and English authors.[5] As with the *Handbook,* the undertaking was huge, and Silvano Arieti was assisted by Marianne, his wife, who helped him correspond with contributors (SAPj).

The first volume appeared in 1971 (Arieti 1971a). Beyond its aim of facilitating scientific communication in psychiatry, it had a theoretical side, as Arieti pointed out in the *Preface*:

> This book ... serves more than a practical purpose. It reflects, as does the restive cultural scene, the new spirit that pervades man. Frontiers, political divisions, and separate histories have increasingly less significance, for while we respect our distinctiveness more and more, we feel more and more alike. This book will, hopefully, prove once more that anyone can benefit from whatever psychiatric achievement is made in any part of the world; it will, thus, in its own way, be a humble testimony to that which needs constant reaffirmation – the universality of man. Further, in that it will simultaneously highlight the benefit that all can derive from that which is specific to certain groups of men in particular conditions in different lands, it will reassert the importance of man's diversity within the greater realm of his universality (Arieti 1971a, p. ix).

Thus, with the *World Biennial* Silvano Arieti reasserted the importance of similarities and differences, universals and particulars, and identity and otherness that he had long put at the core of his clinical and epistemological thought. These were matters that he had been dealing with for more than twenty-five years, ever since his encounter with schizophrenic patients at Pilgrim State Hospital. As he wrote, "every schizophrenic, like every man, is both similar to and different from other patients and men. Here again is that dichotomy, similarity and difference, on which all human understanding is based" (Arieti 1955, p. 5). This concept is a leitmotiv of Arieti's work, and in 1970 was embodied in a new project aimed at affirming both the particular and universal character of psychiatric knowledge.

This venture also expanded from cultural and scientific fields to a political one, as Arieti included the Soviet Union among the participating countries, with a contribution by Andrei Snezhnevky, who had been suggested by Aleksandr Lurija (SAP

[4] The members of the Editorial Board were: Horus Brazil (Brazil); Pierre Pichot (France); Paul Matussek and Gerd Huber (Germany); Julius Zellermayer (Israel); Pier Francesco Galli and Cornelio Fazio (Italy); Masashi Murakami (Japan); Alfonso Millan (Mexico); Carlo Alberto Seguin (Peru); Andrzej Jus (Poland); Andrei V. Snezhnevsky (Soviet Union); Ivan Lopez Ibor (Spain); Gaetano Benedetti and Christian Müller (Switzerland); Silvano Arieti, Francis Braceland, Paul Friedman, Jacques S. Gottlieb, Lothar Kalinowski, Harold Kelman, Stanley Lesse, George Mora, and Jurgen Ruesch (United States).

[5] Among the contributors were Jacob Arlow, Jules Bemporad, Bruno Bettelheim, Pietro Castelnuovo-Tedesco, Robert E. Gould, Leon Roizin, Alberta Szalita, and Yasuhio Taketomo (United States), Carlos P. Barros (Brazil), Peter G.S. Beckett (Ireland), Pierre Deniker (France), Yomishi Kasahara and Kenji Sakamoto (Japan), Einar Kringlen (Norwey), Heinz E. Lehmann and Robertson Unwin (Canada), Francis A. MacNab (Australia), Mara Selvini Palazzoli (Italy), Donald W. Winnicott (England).

1968b, 1968c). During the period of the Cold War, Arieti was gathering authors from both sides of the Iron Curtain. Once again science, with a unified message, proved to be mightier than politics.

On the American home front, the publication of *The World Biennial* coincided with Arieti's political and institutional struggle against the nationalistic restrictions of the American Board of Psychiatry and Neurology, which in 1970 floated the idea of preventing foreign graduates from taking the examination for professional certification. Having learned of this plan, Arieti immediately wrote a sharply critical letter to the most important psychiatric journals:

> I was alarmed, dismayed, and surprised when I read the new rules of the American Board of Psychiatry and Neurology ... I was alarmed because I fear how such discriminatory rules will dishearten those of us who put our faith in America and its noble traditions of equality and justice; dismayed, for we who have begun to teach these foreign graduates and have given them support and approval are put in a difficult moral dilemma. If they are good enough to work in our hospitals, then surely they are good enough to take the Board's examination. I was surprised, because I would not have expected the Board to act in so arbitrary, irrational, and capricious way as it did in its categorical exclusion of foreign graduates ... The idea that a foreign background disqualifies a person from being a good psychiatrist to American patients is disproved by those foreign-trained psychiatrists who do practice in the United States ... History teaches that foreign graduates in America have contributed enormously to psychiatric progress. ... We must therefore conclude that the members of the Board have used the power entrusted to them to make these discriminatory and restrictive rules, rules which do not reflect the interest of psychiatry, are unfair ..., and are contrary to the best traditions of America ... I may be considered by some too personally involved with the victims since I myself would have been prevented from taking the examination in 1944, if such rules existed. I am indeed involved and certainly I identify with the victims. But perhaps this identification permits me to speak with more vigor to all my colleagues and to urge them to raise their voices and to do whatever they can to combat this parochial and unjust regulation (SAP 1970d).

Arieti followed up on his dissent by organizing the Committee Against Discrimination in Psychiatry connected to the American Civil Liberties Union (SAP 1970e, 1971a, 1971b). He sent an ultimatum to Arnold Friedman, President of the American Board of Psychiatry and Neurology, stating that unless there was a public revocation of the new rules regarding eligibility of foreign graduates, the Committee would take legal action against the regulation. The move had an effect, as the American Board immediately rectified its position (SAPh; ABPN 1971).

After the publication of the *World Biennial*'s first volume, Silvano Arieti wasted no time beginning work on the second (Arieti 1973a).[6] He worked very hard, even during his summer vacation in Sardinia. As with the *Handbook,* he had to contend with many headaches—setting deadlines, reviewing chapters, corresponding with

[6] The list contributors included the following names: Julio Aray (Venezuela), José Bleger, Carlos E. Sluzki, and Eliseo Verón (Argentina), Jean Bobon (Belgium), Medard Boss (Switzerland), Léon Chertok (France), Robert B. Davis (India), Dario de Martis (Italy). Akira Fujinawa, Yomishi Kasahara (Japan), Seymour L. Hallek, Turan M. Itil, Arnold M. Ludwig, Saul Rosenthal, and Natalie Shainess (United States), Joseph Marcus (Israel), Alexander Mitscherlich, Margarete Mitscherlich-Nielsen, and Axel Triebel (Germany), Nikola Schipkowensky (Bulgaria), George Vassiliou (Greece), Robert J. Weil (Canada).

authors, arranging translations, and chasing after contributors late with their articles (SAPj). Above all, he was concerned by a changing of the guard at Basic Books, where, in a brief time, two different presidents followed in succession—Jack Lynch and Erwin Glickes (SAP 1972a, 1972e, 1972f, 1972h, 1972j). Compared to his already well-established association with Arthur Rosenthal, his relations with the publishing firm became more complicated. Communication became slower and rarer, as clearly emerges from Arieti's correspondence: "Dear Jack," wrote Arieti to Lynch, "If I were not a psychoanalyst … by now I would be suffering from a deep inferiority complex. I would consider my letters not worthy of being answered and myself not warranting any consideration. In several letters to you I asked you to indicate the date of production and publication of the second volume of the *Biennial*" (SAP 1972a). Arieti was worried that Basic Books was underestimating the value of the *World Biennial,* since in this project his "international reputation [was] at stake" (SAP 1972a). He therefore decided to turn to another publisher for the third volume, to be co-edited with Gerard Chrzanowski. They chose Wiley (SAP 1972b) and changed the title into *New Dimensions in Psychiatry: A World View* (SAPg). The first volume of this new series appeared in 1975, and a second in 1977 (Arieti & Chrzanowski 1975; Arieti & Chrzanowski 1977).

Throughout the decade of the 1970s, Arieti kept working feverishly. He moved his office to 125 East, 84th Street (SAP 1973e; Fig. 5.1), where he spent much of his time treating patients and writing.

Besides the *World Biennial,* he was committed to other projects. He worked on the second edition of the *American Handbook*, which would finally appear in seven

Fig. 5.1 Silvano Arieti and the marble bust of Dante in his office (SAP, Library of Congress, Manuscript Division)

volumes. Called the "*Gone with the Wind* of American psychiatry" (A.G. 1966–1967, p. 1), it was a cyclopean undertaking that caused him many headaches: "I want to sleep at night," he often repeated during its preparation, pleading with contributors to adhere to the schedule (SAPe).

There were, in addition, other books. In 1972 he published *The Will To Be Human* (Arieti 1972a) and, in 1974, the second edition of *Interpretation of Schizophrenia,* which in 1975 won the National Book Award for Science (Arieti 1974a). Other works followed: *Creativity: The Magic Synthesis* (Arieti 1976); *Love Can Be Found*, co-authored with his son James (Arieti & Arieti 1977); *Severe and Mild Depression*, co-authored with Jules Bemporad (Arieti & Bemporad 1978); *The Parnas* (Arieti 1979a); *Understanding and Helping the Schizophrenic* (Arieti 1979b); and many other articles and chapters in journals and compendia of essays.

In coping with this torrent of output Silvano Arieti was aided by his efficient secretary Joan Kirtland, who dealt with his manuscripts and correspondence. "All quiet on the Western front here" (SAP 1977e), she used to reassure her boss on summers, urging him to enjoy the sea and a well-deserved vacation.

With his spirited industriousness Arieti reached the height of success. Acknowledged as a scientist of great depth, he was compared to Jean Piaget in Europe. Not coincidentally, it was to Piaget and Arieti that Benjamin Wolman turned in 1976 to promote the *International Encyclopedia of Psychiatry, Psychology, Psychoanalysis and Neurology*—a colossal work of twelve volumes (SAP 1976d; Wolman 1977).

Arieti was also increasingly asked to evaluate candidates who applied for positions in prestigious universities (SAPa). Yet he himself was not employed by any of those universities. It can be clearly stated that his fame did not correspond to any significant academic power: "I am not under any salary from any institute of any sort," he wrote in 1970 to a colleague in St. Louis (SAP 1970b). In 1973, Walter Bruschi, who had once offered him a directorship in Shreveport, suggested his name for a teaching position at New York University, but the effort went nowhere (SAP 1973b). In 1976 he taught at Columbia University, but for only three months and as an adjunct professor (SAP 1977a). He continued instead to teach (without salary) at New York Medical College, where some Italian students attended his lessons, among whom, besides Marco Bacciagaluppi, were Paolo Decina and Luigi Boscolo (SAP 1972c, 1972k).

Arieti remained "an independent worker" all his life (SAP 1971c). Besides the royalties on his books—which in 1974 Basic Books decided to reduce (SAP 1975a)—his only income was from his private practice. The expenses of his professional trips were not paid for by grants or public funds. He was an independent lecturer, paid and reimbursed by the institutions that invited him (SAPa). Sometimes, in the case of particularly expensive trips, he was compelled to look for a sponsor, as happened on the occasion of the Symposium on the Unconscious planned for 1976 in Tbilisi, USSR, under the auspices of the Georgian Academy of Sciences (SAP 1975c). Arieti tried to find an organization willing to pay at least half of his expenses. He also turned to some pharmaceutical firms, but without success—not surprisingly, as in his search he had proposed to write in return for funds an article or two on the scientific importance of the Symposium, not an attractive offer for the pharmaceutical industry (SAP 1975d).

Arieti's lack of financial and institutional backing was astonishing to those young students who continued, during all the 1970s, to wish to work with him. Even his position at the White Institute had become "a very peripheral one" (SAP 1975b).

"I was very surprised to learn that you have no backing from any institution. But perhaps that is some symptom of the times in which we live," wrote a student from England in 1977 (SAP 1977f).[7] Perhaps she was right. Perhaps, though, something else was at stake, for without administrative and bureaucratic chores, free from any "party line," Arieti could devote himself fully to his work and studies. "I am an isolated worker, and whatever I do, I do independently" (SAP 1976c). Independence was the point. Perhaps his job of autonomous helmsman was not simply thrust upon him; perhaps he chose it.

5.2 To Navigate in Freedom

During the 1970s, the geographical expansion of Arieti's sphere of action was concurrent with a broadening of his thinking on social and cultural issues. This development clearly emerged in the *World Biennial*'s second volume, in the *Preface*, where Arieti wrote that "although the areas of psychiatric interest are more and more numerous, social issues emerge as one of the major focus of interest" (Arieti 1973a, p. vii). He further reaffirmed the importance of social issues in *New Dimensions* (Arieti & Chrzanowski 1977) and in the second edition of the *Handbook*, the second volume of which was devoted to sociocultural and community psychiatry (Arieti 1974–1975).

This course of psychiatry mirrored events that had occurred during the student uprising in France in 1968 and that received impetus from the writings of Michel Foucault, Ronald Laing, and David Cooper (SAP Ub). Nevertheless, in the prevailing anti-psychiatric environment, which tended to consider sociological problems as the prime cause of mental illness, Arieti's voice was original. Rather than focusing on the intersections of mental illness with social conditions—the relevance of which he acknowledged—he turned his attention to the psychological roots of the human crisis in the contemporary world. His thoughts on this subject took shape in 1972 in *The Will To Be Human* (Arieti 1972a), a book "written for a general audience" (SAP 1972d) and translated into Italian in 1978 with the title *Le vicissitudini del volere* (Arieti 1978; Passione 2018).

As pointed out by Jack Bemporad in his review, *The Will To Be Human* reflected a further articulation of Arieti's thoughts on the paradoxes of the human condition:

[7] In her letter, the young English student also wrote, "I have tried very hard to find a department of psychology in any academic institution in this country which teaches psychology the way you do in *The Intrapsychic Self*, but have had no success. I can't understand it. Your ideas seem to bring psychology to life, combining a meticulous attention to the person as a human being with impeccable scientific rigor. Other schools of psychology seem to sacrifice humanity to science,and vice versa".

> Contemporary man finds himself in a paradoxical position. Never before has he had so much power to achieve. His technological capacity to remake nature and himself through genetic engineering is awesome and exhilarating. On the other hand, never before has man felt more vulnerable ... precarious and helpless. It is this dual feeling of omnipotence and helplessness that calls for a theory as to the nature of man and the place and extent of man's freedom. Such a theory, Arieti believes, cannot be aprioristic or based on the slick generalizations which often fill the pages of religious and philosophical works. Rather, such a theory must be based on a careful understanding and analysis of the biological and psychological aspects of man's nature (Bemporad 1973, p. 6).

Firmly objecting to the ideology of determinism, Arieti aimed at promoting a new anthropology based on liberty, hoping that it would contribute to a new "faith in man" (Arieti 1972a, p. 2). In the *Preface* Arieti wrote:

> Obscure social forces, seemingly simple, convincing, and clear, are seeking to endow man with a new image. They portray man as totally conditioned, programmed, and without any role in determining his life. In an age like ours, when many feel that they are drifting aimlessly, or are being used by others, or are too afraid to resist, or too uncertain to choose, too weak to act, it is easy to accept that image. Few are the voices which affirm man's initiative in directing his own steps and building his own home. It is my hope that this book will be one of those voices (Arieti 1972a, p. v).

The reference to conditioning at the beginning of the book reveals Arieti's main target of criticism. The first deterministic ideology to be defeated was the behaviorism of Burrhus F. Skinner, whose idea was to program human beings by sophisticated conditioning techniques. There was "no middle way" between Skinner and Arieti: "the student of psychology must make his choice for the one or the other," as Titchener said of the distance between Wilhelm Wundt and Franz Brentano (Titchener 1929, pp. 3–4). This contrast was recalled and highlighted in the publisher's promotional strategy: "conditioning or freedom? Programming or dignity? Skinner or Arieti?" read an advertising flyer (SAPd).

Arieti's struggle against behaviorism was longstanding. The exclusion of a chapter on behaviorism from the *Handbook,* for example, had been deliberate. Given the pluralistic and inclusive character of that work, its deliberate omission was obviously curious, so much that one reviewer had considered it a political move (Trethowan 1968).

In 1971 Skinner published *Beyond Freedom and Dignity*, in which he invoked a new technology of behavior—a technology capable of solving human problems without the obstacles presented by ethics and values (Skinner 1971). Arieti fiercely criticized this notion, for he believed that removing the ethical dimension from human actions meant reducing "man to a subhuman animal" that is not *beyond* freedom and dignity, but *before* and *below* the experience of freedom and dignity (Arieti 1975a, p. 39).

However, behaviorism was not the only battlefield, as Arieti criticized also those psychiatric conceptions that portrayed human beings as completely programmed by genetic and biological factors and those that depicted them as actuated by inner drives (like orthodox Freudianism) or by the external forces of their social environment (like Neo-Freudianism).

In 1969, with *Love and Will* Rollo May raised his voice against an inclination to determinism in American culture (May 1969). Now, three years later, following the publication of Skinner's *manifesto,* Arieti resumed the debate on free will, fostering an image of human beings as masters of their own will.

Described as a "synthesis of all psychological functions," the will, according to Arieti, is what permits man to make his choices and thereby transcend the restraints of determinism that characterize the world of nature. Freedom depends on will. With these words, Arieti pointed out the close relationship between these two elements:

> We must recognize that total freedom is not available to man. In accepting this limitation man acknowledges his kinship with the rest of the animal kingdom. And even partial freedom is not something he was born with; it is a striving, a purpose, something to be attained. Being born free is a legal concept which, although valid in daily life, has philosophical limitations. Striving for freedom is an unceasing attempt to overcome the conditions of physiochemistry, biology, psychology, and society that affect human life. To the extent that man succeeds in transcending these conditions he becomes self-caused (Arieti 1972a, p. 48).

The idea of total freedom put aside, Arieti insisted instead on the concept of attainable freedom, which he considered a margin "sufficient to change the world, to make history, to cause the rise or fall of man" (Arieti 1975a, p. 40). This margin, however, was also thin and frail, always at risk and needing to be endlessly safeguarded. Its frailty clearly emerged in mental disorders, which showed "how difficult it is for the human being to navigate in freedom" (Arieti 1972a, p. 210). Not by chance, Arieti had begun to deliberate about will in his work with patients, especially his catatonic patients (Arieti 1961), who most heartrendingly showed what happens when a person loses the "most human of his possession—the will" (Arieti 1972a, p. 214).

> If we have some psychiatric disorder, our will is hindered in various way. If we are hysterical, we lose control of some function of our body. If we are phobic, the avoidance of the dreaded event rather than our determination will guide our actions. If we are obsessive-compulsive, we feel obligated to obey internal injunctions, even when they seem absurd. If we are psychopathic we cannot say no to ourselves: we are under the impulse to satisfy our urges immediately. If we are catatonic, we go through a stage in which even our smallest movements can be endowed with cosmic responsibility and guilt; consequently, we do not move at all. Those of us who have intensely studied the catatonic type of schizophrenia consider the catatonic state to be the nadir of the human condition (Arieti 1975a, p. 40).

Although the origins of free will remained undiscovered, according to Arieti it was the emergence of awareness—which required the development of the cerebral cortex—that had enabled human beings to make choices, for with awareness came the possibility of determining actions. Thus, teleological causality—the motivations that characterize the psychological dimension of human life—stands alongside deterministic causality, which rules the natural world. Human beings consciously aim at something in their actions. They desire, choose, and decide (Arieti 1973b).

In his book Arieti described finalism in a developmental framework, as a process ranging from the most basic search for pleasure to the highest levels of ethical motivation. Will could thus be described as "an evolutionary process in the same sense that man in his totality is, not only because of his biology, but also because of his history" (Arieti 1972a, p. 8).

According to Arieti, Carlo Collodi's character Pinocchio, the wooden puppet who becomes a human being as he learns to master and direct his own actions, perfectly symbolized this process. Arieti wrote:

> In a leap of artistic imagination, Collodi bypassed the great evolutionary process of billions of years, which first gave autonomous movement to organic matter, then coordinated motion, finally voluntary acts and moral deeds. Collodi's story created a paradoxical situation: while Pinocchio must act as a moral human being, he is not qualified to do so. As the story evolves he ... acquires a conscience. When this gradual transformation is completed, he deserves to be changed into a child of flesh and blood, a regular member of the human race. But this, let us call it maturation, was not easy (Arieti 1972a, p. 3).

The gradual maturation of Pinocchio's conscience and will was only one of the points of interest of Collodi's story. Another—which could be found also in the Biblical story of Jonah—was that this maturation arose from acts of disobedience. The will, Arieti maintained, starts with a *no*. Referring to developmental psychology he wrote:

> Will starts with a "no", a "no" that the little child is not able to say but is able to enact upon his own body. The child enacts a 'no' when he stops the urine from flowing in spite of the urge to urinate, and the bowels from moving in spite of the urge to defecate. When this possibility of control appears with physio-psychological mechanisms in the human organism, will also appears for the first time ... The individual has a choice. He may allow his organism to respond automatically in a primitive way or not ... It is important to be aware of the fact that ... it is unpleasant to use these inhibitory mechanisms, and the child would not use them if other human beings did not train him to do so. ... Any activity ceases to be just a movement, a physiological function, or a pleasure-seeking mechanism: it acquires a social dimension and thereby becomes an action. Thus, ... a new dimension, the interpersonal, enters. The first enacted "no" is also the first enacted "yes": "yes" to mother, "no" to oneself. ... [But] the child does not accept meekly or obey without protest. ... It is especially from the age of eighteen months to the end of the third year that the child puts up active, willed resistance, as many mothers know very well. ... This stage is called negativistic. The child says 'no' to many suggestions of mother. He does not want to eat, to be dressed, undressed, washed, etc. Mother often has to put up a battle. ... All the "no's" constitute a big "no" to extreme compliance, to the urge to be extremely submissive to the others or to accept the environment immediately. They constitute a big "yes" to the self. ... It is the first spark of that attitude that later on will lead the mature man to fight for his independence or to protect himself from the authoritarian forces which try to engulf him (Arieti 1972a, pp. 18–19).

Continuing his attention toward polarities and contrasts that characterized his thinking from his youth (Chap. 2; SAPf), Arieti emphasized the conflicting forces that shaped the development of will. In the first phase, it seemed contradictory that the initial acts of real volition were not completely independent choices, but acts of obedience—almost as if the will, in its early appearance, denied itself. The same could be said for the subsequent negativistic phase, when "the second decisive *no*" to his mother represented self-assertion—a *yes* of the child to himself (Arieti 1972a, p. 18). Borrowing from the sociologist Talcott Parsons the idea of action as a social fact (Parsons 1951a, 1951b), Arieti explained the link between these two steps of development in an interpersonal framework—arguing that we learn to act intentionally because of our relationship with others: "As Parsons writes, action is concerned not only with the internal structure or processes of the organism, but with the organism as it exists in a sort of relationship" (Arieti 1961, p. 277).

Connecting the origin of voluntary action with interpersonal relations and indicating "the ultimate essence of the human predicament" in the tension between an individual and society, free choice and obedience, autonomy and conformity, Arieti introduced the concept of the relationship between will and authority—an issue Collodi dealt with through Pinocchio's relationship with Geppetto, his father. It was by reuniting with his father in the belly of a whale and reconciling with the very paternal authority (which Pinocchio had contested through multiple acts of disobedience) that the puppet achieved his maturity, developing a conscience and becoming a full human being.

Arieti saw authority as the source of the sense of duty, the origins of which he located in the early stages of an individual's development, when a child introjects his parents' orders and they become his internal moral rudder. The process that Freud described in his concept of Super-Ego the American psychiatrist renamed "endocracy"—a "power which gives orders from inside" (Arieti 1972a, p. 72)—a sort of "psychological precursor to Kant's categorical imperative" (Arieti 1972a, p. 90).

Endocracy is a duplicitous force, however—a two-faced Janus. It has a positive aspect, for without it society itself could not exist: "If men did not have a sense of endocratic oughtness, they would behave like little children or like psychopaths. They would want immediate results, immediate satisfaction of their desires. They would not be able to dedicate their efforts to what they were not concerned with at the present. *Après moi le déluge*" (Arieti 1972a, p. 93). Yet, endocracy also has a negative side, for it is a coercive power capable of greatly decreasing in people "the ability to will or to choose," and undermining their freedom by forcing them to follow categorical imperatives. In its negative aspect, endocracy could be maliciously used for the advantage of a cruel tyrannical power so called—a word that, borrowing from Emil Durkheim's idea of coercive power (Durkheim 1938), Arieti interpreted in its worst meaning:

> The word power ... connotes not a function ... but a force, mostly in a negative sense, ... which is experienced by the individual as thwarting, deflecting, inhibiting or arresting one's will, one's freedom, or one's capacity for growth and expansion. ... Power is ... the capacity to manipulate, control, deflect, exploit, crush the will of others. ... Power affects every interpersonal relation and disturbs it to such a point that a state of communion between two or more people is no longer possible (Arieti 1972b, pp. 16, 22).

An interpersonal framework thus explains the different meanings of "authority" and "power"; whereas "authority" always refers to a relationship that does not hinder a person's development of freedom but supports it with communion and "basic trust" (Buber 1938; Erikson 1950), "power" results in the denial of any dialogical dimension, for it is the imposition of the will of one person on another. Thus *The Will To Be Human* is not only a book about will and freedom, but also a book about the power that hinders will and freedom.

As with other subjects Arieti studied—love and creativity, for example—power cannot not be fully explained by science, or at least, by traditional scientific reductionism. For example, power cannot be explained in biological and naturalistic terms simply as the outcome of the instinct of aggression, as ethological theories suggested (Lorenz 1966), because "at a certain point in the life of man or in the history of the human race, new factors emerge that completely change the aim and potentiality of aggression" (Arieti 1972a, p. 141). In this regard, Freud himself did not go much further when he included aggression in the sphere of a death instinct.

Arieti's point is that power cannot be compared to basic drives or primary needs like hunger, sex, thirst, and so forth. These biological necessities are self-limiting, whereas "some people in search of power cannot conceive any limitation" (Arieti 1972a, p. 139). Unlike biological drives, the need for control and domination can become boundless and self-regenerating.

If the attribution of mere biological causes was not useful for dealing with power, useful hints could instead be found in Alfred Adler's analysis of the feeling of inferiority developed in a child when he finds himself in a world ruled by adults (Adler 1927). Expanding on this idea, Arieti connected this feeling of inferiority to a structural contradiction of the human condition. In particular, taking from the sociologist Alvin Gouldner the concept of "domain assumptions," by which he means those philosophies of life, conscious or unconscious, that shape our personality and direct our actions (Gouldner 1970), Arieti wrote:

> What are the philosophies of life ... which urge people to dominate others? At a certain time in his phylogenetic or ontogenetic development, man transcends his biological nature and becomes aware of a basic irreconcilable dichotomy: he conceives a theoretical or ideal state of perfectibility, and yet he is very imperfect, and in relation to the ideal, inferior. ... He faces a theoretical infinity of space, time, things, and ideas, which he can in a vague sense visualize but not master. On the other hand, he becomes aware of his finitude. He knows he is going to die, and that the range of experiences he is going to have is limited. ... But being able to conceive the infinite ... he cannot accept his littleness. ... He feels frustrated about his own nature and desperately searches for ways to overcome his condition. At a certain period in history some religions have made him conceive compensations in another life, after death. However, these conceptions of immortality were conceived not earlier than 3500 years ago. Earlier in human history the only way to obtain an apparent expansion of the prerogatives of life was to invade the life of others. Since then, this method has remained the prevailing one. My life will be less limited if I take your freedom, if I make you work for me, if I make you submit to me. Thus instead of accepting his limitations and helping himself and his fellowmen within the realm of these limitations, man developed domain assumptions which made him believe that he could bypass his finitude and live more by making other less alive (Arieti 1972b, pp. 24–25).

Moving from a psychological analysis to an ontological one, Arieti recognized at the basis of man's drive for power the paradoxical attempt to overcome his own limited human condition—an attempt that Nietzsche discussed with his philosophy of a "superman." In his book, Arieti sternly criticized the German philosopher, maintaining that instead of wishing to be supermen, "we must will to be human" (Arieti 1972a, p. 253).

Of course, man might have reacted differently to his finitude. For example, he might have accepted it without dismissing his yearning for infinity, a yearning that he might have transformed into a motive for personal and social growth. But history shows, alas, that in place of basic trust and communion, fear and the drive for power generally prevail: "If you do not dominate the other, the other will dominate you" (Arieti 1972b, p. 26).

That fear is a basic ingredient of power is shown, according to Arieti, in the use of terror by tyrants in all generations. In addition to fear, however, tyrannies also use other powerful psychological means, exerting an "endocratic surplus" capable of creating feelings of oughtness in inhuman principles.[8] According to Arieti, an endocracy of this kind characterizes the greatest crimes of history—slavery, colonialism (Fanon 1966),[9] the Armenian genocide, the Holocaust. Along with many other historical tribulations (like Stalinism, Fascism, etc.), these crimes were committed without opposition before a complicit silent majority. Nazi Germany, with its mechanisms of connivance, was an example of this surplus power—as masterfully depicted by Hannah Arendt in her analysis of the "banality of evil" (Arendt 1963). Arieti wrote: "Even Adolf Eichmann during his trial in Jerusalem, invoked Kant in his defense. He said that his duty toward his fatherland transcended any other consideration and justified his crimes" (Arieti 1972a, p. 90). Fascist Italy—whose "abominable king" Victor Emmanuel III was a "typical example of an evil watcher" (Arieti 1972a, p. 183)—could be judged similarly.

Thus the theme of evil—another subject traditionally neglected by science—makes an appearance in *The Will To Be Human.* "The truth," wrote Arieti, "is that psychology and psychiatry have not studied evil as evil. Neither the well-known psychological dictionary by English and English, nor the psychiatric dictionary by Hinsie and Campbell so much as lists the term evil" (Arieti 1972a, p. 224; English & English 1958; Hinsie & Campbell 1960). This sentence was a wake-up call to psychiatry. It was time for science to deal with the subject, Arieti exhorted, for it would be wrong to consign the historical abuses of power to the past. The endocratic surplus and the silent majority that does not oppose evil but actively participates in it is a characteristic of our modern democracies as well. Just think, he suggested, of the followers of Charles Manson who followed the orders of their charismatic leader to commit a massacre on Cielo Drive.

According to Arieti, however, the eliciting of endocracy was not restricted to such extraordinary cases; it was much more common:

> The majority of people in some democratic countries have become so unaware of basic political realities that they don't seem to recognize the existence of unfair accumulations of power. ... The 150 largest firms in the United States produce more than half of the country's manufactured goods and control a large segment of the country's economy. ... Thus power concentrates in a few hands to a much greater extent than people realize. Not only do corporate managers establish the prices that they believe the market will bear with little interference from the government; they mold the taste of people to fit their own purposes; they constantly

[8] About the concept of endocratic surplus Arieti pointed out that, although the word surplus was often connected with Marxian theory, in his book it had nothing to do with Marx and was not intended as an economic concept, but as a psychological one.

[9] On colonialism Arieti referred to Fanon's book entitled *Les Damnés de la terre,* published in France in 1961 and translated in English in 1966.

> promote in people a need for new products ... The individual ... now is under a constant persuasion to *buy, buy, buy! consume! consume! consume!* (Arieti 1972a, pp. 184–185).

Thus, in his study of the psychological mechanisms of power, Arieti compared two seemingly unrelated phenomena that he considered as different facets of the impending crisis of Western civilization.

In his analysis Arieti paid attention also to another facet of power, the institutional endocracy of academic life, where independent research is often sacrificed for a dialectic of power that secures a tacit bond between a chairman and his collaborators, who are later to be rewarded for their services with academic titles. This mechanism, by the way, also included the so-called independent institutions. Referring to an exposé published by Erich Fromm in 1958 (Fromm 1963), Arieti wrote:

> Alas! We may avoid Scylla and veer toward Charybdis. In a brilliant article published in 1958 in the *Saturday Review,* Erich Fromm exposed the intrigues which had taken place in the psychoanalytic association in its early history. By interviewing an quoting witnesses, Fromm documented the historical truth of these power manipulations. We would be extremely naïve to think that these maneuvers go on only among the orthodox Freudians. Let us face it! Power games are played all over (Arieti 1972b, p. 31).

Arieti's words can be easily recognized as an echo of his own experiences. In a lecture he gave in 1971 he had already expressed his thinking on this subject, and his friend and colleague Ian Alger, who was present at the lecture, immediately highlighted the biographical relevance of Arieti's views:

> His sincerity is apparent because in the very act of presenting this challenging paper he has demonstrated his own courage in delineating so explicitly the corrupt intricacies of the power structures which he personally has been able to observe and to experience in psychiatric hospitals, psychiatric departments, and psychoanalytic institutes ... Because his own power base is small, by telling it like it is, he runs the risk of personal loss, and also the risk ... that anyone who so speaks is in danger of being labeled paranoid. My own experiences give corroborative evidence that the power structures and power manipulations he has described do indeed exist not only in the institutions he has mentioned ... It would have been tempting, and it would have been much easier, had Dr. Arieti spoken in generalities, and left us comfortable with our suppressions undisturbed. Instead, by his examples ... he has created a more urgent and immediate personal reason for us as analysts to utilize the understanding of power ... not only to bring our newer understandings ... to the therapeutic relationship, but also to take the risk now, for ourselves, in confronting more directly the external hostile forces. He himself has personally taken that kind of risk. Those of us who are staff and faculty members of psychiatric and psychoanalytic hospitals and departments can accept the challenge and also take the risk of immediately moving to identify and oppose the power of unjust authoritarianism in these systems (SAP 1971d).

Beyond biography, and from a professional point of view, there is no doubt that, by addressing the dialectic between freedom and power, Arieti dealt with a problem inherent in psychiatry's identity. To begin with, that dialectic concerned therapy—first, in mental hospitals, where often the director "acts like a king who rules over the life not only of patients but also of the medical and paramedical staff" (Arieti 1972b, p. 28), and then, in analytical settings, where therapy can turn into repression when the therapist aims at conditioning and controlling the patient rather than at enhancing his possible choices (Arieti 1972b; SAP 1974d).

In short, Arieti suggested, whatever the approach or technique, the psychiatrist must always bear in mind that he plays a role on the thin line between knowledge and power—a call to awareness that resembles the sense of a struggle undertaken in those years in Italy by a combative and committed generation of dissident psychiatrists (Babini 2009, 2014; Foot 2015).

Beyond references connected to his own professional career, in *The Will To Be Human* Arieti took inspiration also from his personal and private life. Indeed, the book is full of autobiographical memories that merge with his theorizing. His youthful enthusiasm for Mussolini and the resulting conflict with his father, for example, match up with his analysis of power and authority. The endocratic persuasiveness of Fascist indoctrination during his school years contrasts with his father's call to independent thinking, for "will starts with a *no*"—the *no* to the Duce and his infamous laws that Italians were not able to say.

As conveyed by the book, Arieti's biography reminds readers of Pinocchio and Jonah: like them, he experienced a shipwreck, was swallowed by a whale, and after long years of suffering away from home finally reunited with his father and became a man. In this regard, *The Will To Be Human* is also a book of a witness, for Arieti had personally experienced the brutality of an external power that nullifies freedom. Afterwards, he had not allowed himself "to become a supporter or a silent watcher of the oppressor," and had chosen to strive with all his force toward increasing autonomy. He succeeded in this effort personally and professionally, as he found his way in the New World. The same choice, he wrote, is open to all:

> By sharpening his eye and maintaining a state of alertness the citizen can recognize in the areas of society in which he is involved many situations in which power accumulates to the detriment of the many and to the nonspiritual aggrandizement of the few. It is part of the exercising of one's will to discover these intricate maneuvers and to oppose them to the best of one's ability (Arieti 1972a, p. 188).

In short, that "faith in man" about which he wrote at the beginning of his book seems to be at one with his faith in himself—that is, in the capacity to choose.

5.3 Sailing Against the Tide

The analysis of power in *The Will To Be Human* has roots not only in Arieti's professional and personal experience, but also in the social and cultural context of those years, characterized by the discussion of freedom and power in student uprisings "on various campus in America and Europe" (Arieti 1972a, p. 122). The book must therefore be read in the context of the ongoing student protests and the meaning the protests reflect—the search for new areas of freedom for young people (Meucci 1978).

Nevertheless, Silvano Arieti did not entirely embrace the culture of dissent of those years. For this reason, *The Will To Be Human* can be considered an atypical "militant" book; on the one hand, it values protest movements for their struggle against the endocratic surplus, since "not only persons, but also ideas, institutions, and traditions engender endocracy in the psyche of the individual" (Arieti 1972a, p. 125). On the other hand, it harshly criticizes some claims of young dissenters, on the grounds that they would result in only an ephemeral emancipation and superficial solution to the problem of power.

According to Arieti, there were different sides of the protest movement: it had a complex appearance that escaped the analysis of the American historian Theodore Roszack, the very scholar who had coined the term "counter culture" (Roszak 1965). Roszak, Arieti believed, failed to realize that within the counter-culture movements there were different groups with different goals and had mixed them together in an all-embracing category of "the Great Refusal."

Kenneth Keniston, a social psychologist, focused instead on the different attitudes of dissenters along a *continuum* ranging from alienation to commitment (Keniston 1968a, 1968b, 1971). According to Keniston, who was inspired by the work of Erik Erikson (Erikson 1963, 1968), committed and alienated youth could be distinguished according to their differing levels of the maturation of their self-identity, for some were characterized by a sound self-awareness, while others, who had lost contact with their inner selves, had given up the quest for their personal authenticity and dealt with their surroundings by means of different attitudes—passive adjustment, estrangement, and so on.

Taking a cue from Keniston's analysis (SAP 1972p),[10] Arieti added another group of young people to committed and alienated youth, for whom he coined the name "Pinocchians":

> One of the basic tenets of the Pinocchian philosophy can be summarized in two short sentences: "Be. Do not become." In other words, spend your time in being and enjoying the way you are now, and not in trying to become something else tomorrow. Live to the fullest possible extent now. ... Whoever tells you to direct your life in accordance with what your future demands of you is to be distrusted. ... The subculture teaches "not to wait." ... We must be fully involved with the present moment.
>
> But intense involvement with the present does not imply that we should give up "becoming" altogether. As a matter of fact, it is mainly in the process of becoming that our human nature distinguishes itself from that of animals. A cat, a dog, a horse change very little throughout their lives. They ... have practically no symbolic functions and therefore must remain within the limits of what is immediately offered to their senses. ... Man, on the other hand, is always becoming because he is constantly affected by a universe of symbols. Today he is somewhat different from the way he was yesterday, and tomorrow he will be different from the way he is today. He listens, reads, learns ... There is no definite end to the possible growth of man. ... No mature will is possible without the ability to postpone. Only an ability to overrule the present makes us capable of making choices and of "becoming." Giving up becoming means returning to an immature stage of will (Arieti 1972a, pp. 101–102).

According to Arieti, the emancipation the Pinocchians desired was illusory. By giving up becoming and refusing to make choices, they actually dismissed the most human of their faculties—the will. Unlike Collodi's puppet, who turned into a human being by *acting* disobediently, Pinocchians were trapped in the Land of Toys, where

[10] In 1972, at the seventeenth Annual Meeting of the American Academy of Psychoanalysis, Arieti met Keniston and discussed his books.

they abandoned the road to maturity and unleashed their Id. They wallowed in the belly of the whale without rejoining Geppetto.

Arieti's analysis of these young protesters was not gentle. He considered their rejection of culture and traditions unacceptable. "Undoubtedly there are *many* things that we should refuse in our culture," he cautioned, "but let us not put our whole cultural heritage into one wastebasket" (Arieti 1972a, p. 108). Above all, he denied that emancipation could stem from an instinctual frantic pleasure-seeking fulfillment rather than from learning and reasoning. Thus, in addition to his criticism of Skinner's ideas, Arieti was critical of Freud's conception of man's instinctual nature. According to him, deterministic assumptions underlay both behaviorism and psychoanalysis (Mora 1973).

On this subject Arieti's book echoes the thinking of French psychoanalyst Jacques Lacan on the "evaporation of the father" (Lacan 2017; Recalcati 2019). In 1969, in a seminar at the Sorbonne, Lacan had discussed power by focusing on the conflict between instinctual satisfaction and civilization, following the path of Freud in *Totem and Taboo* and *Civilization and Its Discontents* (Lacan 1991; Roudinesco 1997). Though we know that Arieti read Lacan's *Écrits* (Lacan 1966; SAP 1969e), no reference to the French psychoanalyst appears in the bibliography of *The Will To Be Human*. Arieti's exchange with him seems to have occurred as a scholarly "dialogue-at-a-distance"—an unexpected convergence of authors from different cultural traditions in an age of global protest and dissent that stimulated a common deliberation on power, emancipation, and the role of psychoanalysis.

Like Lacan, Arieti discussed Freud's views about the relationship between innate drives and reason, instincts and civilization. In *The Will To Be Human* he wrote:

> More than any other predecessor or contemporary [Freud] revealed the role of the unconscious wish as a determinant of human behavior. But this great contribution does not imply that will has no role to play at all. Let us remember another one of Freud's famous statements: "Where id was, ego must be." The id is the agency of the psyche where most of the unconscious wishes are located. As a result of psychoanalysis, Freud thought that the unconscious wishes must become part of the ego; that is, become conscious. But what is the purpose of their becoming conscious? To make the id part of the ego does not mean to unleash the unconscious wishes, to give them free reign or unrestricted access to conscious behavior, although this is the way in which some people prefer to interpret Freud's statement. ... Psychoanalysis, as conceived by Freud, is no liberation of the id – in the sense of unleashing primitive unconscious wishes – but is a liberation of the ego from the unwanted, unconsciously determined, oppression exerted by the id. Thus, *psychoanalysis has the function not of restricting, but of enlarging the sphere of influence of the will* (Arieti 1972a, p. 44).

Thus, Arieti believed that psychoanalysis could contribute to human emancipation, on the condition that it was conceived as a gateway to *logos* and reason rather than to instinctual drives.

In his book Arieti asserted this same view also in regard to sexual liberation. Referring to Wilhelm Reich and Herbert Marcuse (Reich 1949; Marcuse 1955), he wrote:

> Many young people have found inspiration and support in the writings of Reich, Marcuse, and other writers with similar views in their special interpretation of Freud. These people view

> psychoanalysis as conductive to a philosophy of life which permits unrestricted sexuality and uninhibited sensuality. Hedonism finally acquires respectability. What Aristippus and Epicurus were not able to do, Freud, Reich, and Marcuse would have accomplished: the unification of will with the primitive wish – and not an incomplete immature will, similar to the impulse-response mechanism of the animal, but a more complicated mechanism by means of which "advanced" human beings remove the distance between stimulus and pleasant response (Arieti 1972a, p. 59).

Despite acknowledging Reich's and Marcuse intellects, Arieti did not agree with a common interpretation of their writings, which he saw as describing man "as a fundamentally hypothalamic organism" (Arieti 1972a, p. 62). In particular, he identified a confusion that formed the basis of the thinking of Reich and Marcuse—the failure to articulate the difference between wishes and drives. According to Arieti, this confusion was rooted in the ambiguity of the Freudian concept of wish, which implied two different meanings: on the one hand, in explaining psychological phenomena in physical terms, Freud conceived a wish as a *force*—"like a charge of electrical energy which gives man the power to act" (Arieti 1972a, p. 35). On the other, according to Freud, a wish was "a mental representation of something that, once attained, would give pleasure"—that is, as a mental activity entailing a symbolic function, and therefore "much more complicated than a force or an appetite" (Arieti 1972a, pp. 34–35). As far as sexuality was involved, however, Arieti believed that the word "wish" should be defined as a symbolic function. "In human beings, just as aggression becomes associated with symbolic processes which may lead to hate, sexuality becomes associated with symbolic processes which may lead to love" (Arieti 1972a, p. 64). Thus, another topic that could not be analyzed by traditional scientific methods appeared in *The Will To Be Human*—love, a subject to which Silvano Arieti devoted another book, written with his son James and published in 1977 (Arieti & Arieti 1977).

This book did not have the success that it might have deserved, though (SAPb). Although written for a lay audience, it did not have the same "appeal of the psychological how-to books" that had invaded the market—books that, according to Arieti, did "more harm than good" (SAP 1977b). Above all, in those years *Love Can Be Found* appeared as a maverick book; it was different from the many that crowded the bookstore shelves, for it avoided a "gymnastic approach" to sex. "In Victorian times," one reads in a review, "they complained the common malaise was because anything to do with sex was never discussed. Today, everything about sex is discussed endlessly, and there is still a terrible malaise … and that's not due to lack of sex, but to lack of love" (Allison 1977, p. 48).

The point of view conveyed by Silvano and James Arieti's book had already been presented in *The Will To Be Human.* Despite being written in an era marked by a general demand for the greatest possible sexual freedom, the book had suggested a view of sex that seemed conservative to many. In 1973, for example, when Silvano Arieti published in *Psychology Today* the excerpts from *The Will To Be Human* that were about Reich and Marcuse (Arieti 1973c), the magazine was flooded with letters of protest that accused him of moralizing (SAPa). "It is apparent," one of these letters read, "that Reich and Marcuse have hit Arieti where it hurts" (SAP 1973a).

To Arieti's defense promptly came Natalie Shainess, a well-known psychiatrist who had written a chapter on the "Psychological Problems Associated with Motherhood" for the third volume of the *American Handbook of Psychiatry.* Endorsing Arieti's criticism of the hedonistic and mechanical concept of sexuality (SAP 1973g), she wrote to *Psychology Today:*

> I want to express my admiration of Dr. Silvano Arieti's article ... As a colleague of his, I have had ample opportunity to notice the clarity and profundity of his thinking, in other areas, as well as the sexual. I can anticipate, having also written about sex from a similar vantage point, that there may be many who will criticize and attack his thinking. He needs no defense, and yet he should have some. ... In life's course we must make choices. If we gain one thing, we lose something else. Who can prove that the views of Reich and Marcuse will guarantee greater satisfaction in life? It will not. True sexual pleasure is related to the quality of the person and the relationship, not just of the body. Throwing off all restraints and using the sexual apparatus incessantly and with anyone, is not the means of achieving greatest sexual satisfaction. ... Another way of conceptualizing the "peak experience" is that of "shared intimacy." This requires the sharing of mind, body values and emotions. It also comes from a shared learning process, sexual and other. ... Compulsive sex is not free sex, and a belief that one must at all times be sexually active regardless of circumstance, is not sexual freedom. ... *Bravo* to Dr. Arieti for daring to go against the tide (SAP 1973d).

What is certain is that Arieti's point of view ran counter to "some of the most powerful cliches" (Mora 1973, p. 939) of those years. His thinking was a world apart from the idea of "anarchical desire" that was at the heart of Deleuze and Guattari's *Anti-Oedipus*, a classic reading for many European protesters (Deleuze & Guattari 1972; Dosse 2010). Nor did Arieti's ideas of emancipation and freedom accord with other traits present in American culture, like those conveyed by the hippie movement, which he considered the expression of a new form of conformism, a view reminiscent of Italo Calvino's sharp analysis of the beatnik youth, according to which the perfect efficiency of a conformist society is revealed by its ability to create a corps of nonconformists in the uniform of a nonconformist (Calvino 2002).

According to Arieti, Pinocchians' conformism clearly emerged also in their customary condemnation of science, despite the desire to benefit from its results. In this way, the desire of Pinocchians to profit from technological goods was different from the wiser critical attitude of the committed young people who cautioned "against those results of science which, by being wrongly applied, contribute to a total objectification of life, consumerism, and loss of basic values" (Arieti 1972a, p. 117). Here too, the choice of what man decided to do with his knowledge was critical:

> Scientific knowledge and control of nature do not necessarily lead to evil results. They may lead to the conquest of polio as well as to the construction of gas chambers, to man landing on the moon or to the dropping of the bomb on Hiroshima. Whether scientific methodology will be used for good or evil is determined not by science itself but by the use men (scientists and nonscientists alike) make of it. In our twentieth century reason and science have been used on the side of evil, and results such as penicillin, polio vaccine, and reaching the moon, are pale in comparison to such catastrophes as two world wars, the murder of six million Jews by the Germans, and the use of the atom bomb. What was at fault in these cases was not science itself, but the way science was used (Arieti 1972a, pp. 102–103).

Thus, according to Arieti, it was not among Pinocchians that the solution of social problems was to be found. Pinocchians seemed to embody William Golding's parable in *Lord of the Flies*, where the forces of Id, in all their virulence, take over in the self-government of immature children marooned on a desert island until adults suddenly arrive and restore order (Golding 1954). Golding's 1954 book achieved great success; "I wonder," Arieti wrote, "whether the book would have had equal success … had it been published in the late sixties, when the craving for id-liberation became stronger and stronger" (Arieti 1972a, p. 109).

Arieti's assessment of "newly committed" youth was quite different, for, as he said, they represented the better voice of the protest movement. He welcomed the young people who struggled against excessive endocracy and petrified principles as initiators of a new project of human community. They did not seek happiness by an unrestrained Id, like Pinocchians, but by their commitment to affirm the ethical aim of existence.

Nevertheless, Arieti was aware that the concept of ethical aim was difficult to define clearly. He knew, for example, that the Ten Commandments and the other laws in the Torah were not sufficient, since in human hands laws and commandments were subjected to significant, even atrocious, distortions. Although laws should always be inspired by ethics, ethics could not be equated with laws.

We must recognize that in *The Will To Be Human* some of Arieti's thoughts on ethics seem here and there to take a dead end of platitudes: "Nature should be transcended and human history should become ethical history." "In ethical progress mankind should find its greatest source of dignity." To be free means "to make use of freedom by following the ethical way." "The essential activity of man is the good deed." And so on. Neither is Arieti helped by references to Plato, since "his conceptions were entirely ideal, and had no practical feasibility," or by references to Hegel and Marx, who identified the aim of ethics, respectively, in the supremacy of thought and work: "It seems to me," Arieti wrote, "that thought and work must have an ethical quality, and must become ethical events" (Arieti 1972a, pp. 225–228). This statement, although suggestive of his idealistic streak, does not contribute to a clear definition of his conception of ethics.

A more useful understanding can be gained if we turn to Arieti's psychiatric thinking as presented in 1971 at the International Symposium on Psychotherapy in Finland and then restated in his book (SAP 1971h; Arieti 1971b). At the Symposium, Arieti dealt with the views of Finnish psychiatrist Martti Siirala, a pupil of Gaetano Benedetti (SAP 1972q). Siirala had described the schizophrenic as a victim and as a prophet to whom nobody listens. His prophecies consisted of "insights into our collective sickness, into the murders that we have committed for many generations and which we have buried so that they will not be noticed" (Arieti 1972a, p. 193). By showing us the sick face of our communities, Siirala suggested, the schizophrenic, by means of his dissident delusions, was carrying out a revolutionary action (Siirala 1963).

Though Arieti disagreed with his Finnish colleague about identifying mentally ill persons with revolutionaries, he acknowledged the ethical value of the schizophrenic's existential condition—a condition that he compared to that of the Old Testament's prophets (Arieti 1974b). In his book he wrote:

> I believe that the schizophrenic ... cannot be literally called a ... prophet, an innovator, a dissenter, a revolutionary. Nevertheless, we must acknowledge that the schizophrenic responds more to universal hostility than we do, and crumbles more than we do. The psychological pollution of the world seems to concentrate its effects on him ... . He reminds me of those fish which absorb mercury discharged into the ocean; those fish tell us that mercury should not be discharged into the ocean. ... Although the schizophrenic exaggerates and greatly distorts the world's hostility, he reminds us that such hostility exists ... He may help us to become aware that our "normality" requires mental mechanisms whose validity is questionable. Normality presupposes adaptation or capacity for adjustment. At times what is required on our part is callousness toward harmful stimuli. We protect ourselves by denying them, hiding them, becoming insensitive, or finding a thousand ways of rationalizing them or adjusting to them. We become the "silent majority." ... By being so sensitive and so vulnerable the patient counteracts the callousness that our "healthy" capacity for adjustment and adaptation has brought about. His message is, "Do not compromise at the cost of mutilating your soul. Do not become the ally of the hostile powers. Do not become a watcher of evil." Thus, there are some values that we can share with the schizophrenic ... The basic value of the schizophrenic is actually the basic value of the human being. He wants to be the sovereign of his will, to be totally himself. He does not succeed. As a matter of fact, he finds sovereigns all over, but not in himself. ... He feels other people determine that the world will be as it is, but that he has no part in that determination (Arieti 1972a, pp. 196–197).

Criticizing the conversion by social Darwinists of the biological ideas of the struggle for life and natural selection into moral and social values, Arieti described the schizophrenic as a person completely unfit for any type of jungle. With his vulnerability the schizophrenic reminds us of the importance of brotherhood in our human community. In short, his condition suggests the need for an ethics conceived not as pre-established law but as a conscious and responsible sharing of our world. Thus, Arieti's thought turned into a "universalist" conception of life and a communal idea of ethics—conceptions that he would soon reiterate in other writings.

The Will To Be Human was published in September 1972 by Quadrangle Books, a subsidiary of *The New York Times.* Arieti's choice of this firm reflected a wish to write for a wider audience of lay readers and to assert himself as a writer in addition to being a psychiatrist. With this goal in mind, in the 1970s he also turned to writing stories and plays, for some of which he used the pseudonym Abramo Tuscany (SAPc; Arieti 2016). "How you get time to do these things, how you ever get them done, I don't know, but I am in open admiration," wrote Francis Braceland in October 1972 (SAP 1972l).

Considered by Gaetano Benedetti as one of the major psychiatric contributions to the issues of our time (SAP 1972o) and endorsed by Bruno Bettelheim,[11] *The Will To Be Human* was nominated for the National Book Award in Philosophy and Religion. It was included among the finalists, but did not win (SAPd; SAP 1973f).

Though disappointed—as he felt that it was the best book he had written (SAP 1972n)—Arieti was not demoralized and did his best to generate interest in the

[11] Bettelheim's endorsement appeared on the jacket's cover: "Dr. Arieti deals with one of the most important and troubling problems of today, and his contribution to our understanding of the role of will in human affairs is most penetrating".

book among lay readers and among his colleagues in the United States and abroad (SAP 1972i), hoping it would stimulate a discussion on ethics and on the social responsibilities of psychiatrists (SAPd). With this purpose in mind, he also supported the candidacy of Louis Joylon West (SAP 1973h, 1973i, 1973j, 1974a, 1974b), a psychiatrist committed to civil rights, to lead the American Psychiatric Association (Barton 1987).

Above all, with *The Will To Be Human* Arieti aimed at revitalizing Sullivan's conception of psychiatry as "a science not only qualified to deal with the mentally ill, but a science having vast relevance in human affairs, touching … many fields of human endeavor and human problems" (Sullivan 1955, p. 10). "A psychiatrist cannot help but venture beyond psychiatry" (Arieti 1975b, p. 1, my translation), Arieti added. In this respect, *The Will To Be Human* ran doubly against the tide—both for its cultural criticisms and its humanistic and comprehensive conception of psychiatry at a time when psychiatry was narrowing its focus on biological factors.

References

ABPN (1971). Briefly Noted. ABPN Notice. *Psychiatric News,* 6 (4), 8.

A.G. (1966–67). The American Handbook of Psychiatry. *William Alanson White Newsletter,* 1 (2), 1.

Adler, A. (1927). *Understanding Human Nature.* Garden City Pub.

Allison, J. (1977, July 2). Father, Son, Present Love Above Erotica. *Indianapolis News,* 48.

Arendt, H. (1963). *Eichmann in Jerusalem. A Report of the Banality of Evil.* Viking Press.

Arieti, J. (2001). Ricordi del figlio di uno psichiatra. In R. Bruschi (Ed.), *Uno psichiatra tra due culture. Silvano Arieti 1914–1981* (pp. 29–37). ETS.

Arieti, J. (2016). The Literary Ambitions and Storytelling Art of Silvano Arieti. *Academy Forum,* 60, 11-14.

Arieti, S. (1955). *Interpretation of Schizophrenia.* Basic Books.

Arieti, S. (1961). Volition and Value: A Study Based On Catatonic Schizophrenia. *Comprehensive Psychiatry,* 2 (2), 74-82.

Arieti, S. (Ed.) (1969–1970). *Manuale di psichiatria.* Boringhieri.

Arieti, S. (Ed.). (1971a). *The World Biennial of Psychiatry and Psychotherapy. Vol. I.* Basic Books.

Arieti, S. (1971b). Psychodynamic Search of Common Values with the Schizophrenic. In *International Congress Series N. 259. Psychotherapy of Schizophrenia. Proceedings of the IV International Symposium, Turku, Finland, August 4–7* (pp. 94–100). Excerpta Medica.

Arieti, S. (1972a). *The Will To Be Human.* Quadrangle.

Arieti, S. (1972b). The Origin and Effect of Power. In J. H. Masserman (Ed.), *Science and Psychoanalysis. Vol. XX: The Dynamics of Power* (pp. 16–32). Grune & Stratton.

Arieti, S. (Ed.). (1973a). *The World Biennial of Psychiatry and Psychotherapy. Vol. II.* Basic Books.

Arieti, S. (1973b). Causality, Awareness, and Psychological Events. In W. Gray & N.D. Rizzo (Eds.), *Unity Through Diversity: A Fetschrift für Ludwig von Bertalanffy* (pp. 843-869) Gordon & Breach.

Arieti, S. (1973c). Marcuse and Reich Are Wrong About Sex. *Psychology Today,* 6 (8), 26–30, 88.

Arieti, S. (1974a). *Interpretation of Schizophrenia.* Second Edition. Basic Books.

Arieti, S. (1974b). An Overview of Schizophrenia from A Predominantly Psychological Approach. *The American Journal of Psychiatry,* 131 (3), 241-248.

Arieti, S. (Ed.). (1974–1975). *American Handbook of Psychiatry.* Second Edition. Basic Books.

Arieti, S. (1975a). Psychiatric Controversy: Man's Ethical Dimension. *The American Journal of Psychiatry,* 132 (1), 39-42.

Arieti, S. (1975b). *Psichiatria e Oltre.* Il Pensiero Scientifico.
Arieti, S. (1976). *Creativity: The Magic Synthesis.* Basic Books.
Arieti, S. (1978). *Le vicissitudini del volere.* Il Pensiero Scientifico.
Arieti, S. (1979a). *The Parnas.* Basic Books.
Arieti, S. (1979b). *Understanding and Helping the Schizophrenic.* Basic Books.
Arieti, S., Arieti, J. (1977). *Love Can Be Found: A Guide to the Most Desired and Most Elusive Emotion.* Harcourt Brace Jovanovich.
Arieti, S., Bemporad, J. (1978). *Severe and Mild Depression: The Psychotherapeutic Approach.* Basic Books.
Arieti, S., Chrzanowski, G. (Eds.). (1975). *New Dimensions in Psychiatry: A World View. Vol. I.* John Wiley.
Arieti, S., Chrzanowski, G. (Eds.). (1977). *New Dimensions in Psychiatry: A World View. Vol. II.* John Wiley.
Babini, V. P. (2009). *Liberi tutti. Manicomi e psichiatri in Italia: una storia del Novecento.* Il Mulino.
Babini, V. P. (2014). Looking Back: Italian Psychiatry from Its Origins to Law 180 of 1978. *The Journal of Nervous and Mental Disease,* 202 (6), 428-431.
Barton, W. E. (1987). *The History and Influence of the American Psychiatric Association.* American Psychiatric Press Inc.
Bemporad, J. (1973). The Will To Be Human. By Silvano Arieti. *Reform Judaism,* 1 (8), 6.
Buber, M. (1938). *I and Thou.* T. & T. Clark.
Calvino, I. (2002). *Un ottimista in America 1959–1960.* Mondadori.
De Risio, A. (2019). *The Italian Psychiatric Experience.* Cambridge Scholar Publishing.
Deleuze, G., Guattari, F. (1972). *Capitalisme et Schizophrènie, L'Anti-Œdipe,* Minuit.
Dosse, F. (2010). *Gilles Deleuze & Félix Guattari: Intersecting Lives.* Columbia University Press.
Durkheim, É. (1938). *The Rules of Sociological Method.* The Free Press.
English, H. B., English, A. C. (1958). *Comprehensive Dictionary of Psychological and Psychoanalytical Terms.* Longmans Gree.
Erikson, E. (1950). Growth and Crisis of the Healthy Personality. In M.J. Senn (Ed.), *Symposium on the Healthy Personality* (pp. 91-146). Josiah Macy Jr. Foundation.
Erikson, E. (1963). *Youth: Change and Challenge.* Basic Books.
Erikson, E. (1968). *Identity: Youth and Crisis.* Norton & C.
Fanon, F. (1966). *The Wretched of the Hearth.* Grove Press.
Foot, J. (2015). *The Man Who Closed the Asylums: Franco Basaglia and the Revolution in Mental Health Care.* Verso.
Fromm, E. (1963). Psychoanalysis: Science or Party Line? In *The Dogma of Christ and Other Essays on Religion, Psychology and Culture* (pp. 131–144). Holt, Rinehart and Winstson.
Golding, W. (1954). *Lord of the Flies.* Penguin Books.
Gouldner, A. (1970). *The Coming Crisis of Western Sociology.* Basic Books.
Hinsie, L. E., Campbell, R. J. (1960). *Psychiatric Dictionary.* Oxford University Press.
Keniston, K. (1968a). *The Uncommitted: Alienated Youth in American Society.* Harcourt Brace.
Keniston, K. (1968b). *Young Radicals: Notes on Committed Youth.* Harcourt Brace.
Keniston, K. (1971). *Youth and Dissent: the Rise of a New Opposition.* Harcourt Brace.
Lacan, J. (1966). *Écrits.* Seuil.
Lacan, J. (1991). *L'envers de la psychanalyse. Seminaire XVII, 1969–1970.* Le Seuil.
Lacan, J. (2017). Note on the Father and Universalism (1968). *The Lacanian Review,* 3, 10-11.
Lorenz, K. (1966). *On Aggression.* Harcourt.
Marcuse, H. (1955). *Eros and Civilization: A Philosophical Inquiry Into Freud.* Beacon.
Masters, W. H., & Johnson, V. E. (1966). *Human Sexual Response.* Little Brown.
May, R. (1969). *Love and Will.* Norton & Company.
Meucci, G. P. (1978). Un colloquio come presentazione. In S. Arieti, *Le vicissitudini del volere* (pp. XI–XIX). Il Pensiero Scientifico.
Mora, G. (1973). Book Review. Arieti, Silvano, M.D.: The Will To Be Human. New York, Quadrangle, 1972. *Bullettin of the New York Academy of Medicine,* 49 (10), 938–940.

Parsons, T. (1951a). *Toward a General Theory of Action.* Harvard University Press.
Parsons, T. (1951b). *The Social System.* The Free Press.
Passione, R. (2018). Language and Psychiatry: the Contribution of Silvano Arieti Between Biography and Cultural History. *European Yearbook of the History of Psychology,* 4, 11-36.
Recalcati, M. (2019). *The Telemachus Complex: Parents and Children After the Decline of the Father.* Polity Press.
Reich, W. (1949). *Character Analysis.* Orgone Institute Press.
Roszak, T. (1965). *The Making of a Counter Culture: Reflections on the Technocratic Society and Its Youthful Opposition.* Doubleday.
Roudinesco, E. (1997). *Jacques Lacan.* Columbia University Press.
SAPa. *Correspondence, 1940–1981.*
SAPb. *Speeches and Writings, 1940–1981. Love Can Be Found. Promotion and reviews, 1976–1977.*
SAPc. *Speeches and Writings, 1940–1981. Plays.*
SAPd. *Speeches and Writings, 1940–1981. Will To Be Human. Promotion, 1972.*
SAPe. *Subject File, 1914–1981. American Handbook of Psychiatry, editor. Correspondence, 1965–1980.*
SAPf. *Subject File, 1914–1981. Education. Workbooks, circa 1930.*
SAPg. *Subject File, 1914–1981. New Dimensions in Psychiatry, 1976.*
SAPh. *Subject File, 1914–1981. Organization. American Board of Psychiatry and Neurology.*
SAPi. *Subject File, 1914–1981. Travel Planning, 1959–1980.*
SAPj. *Subject File, 1914–1981. World Biennial of Psychiatry and Psychotherapy, editor. Correspondence.*
SAP (1960). L, April 26 (Stanley Lesse to Saul Feldman). In *Subject File, 1914–1981. Travel Planning, 1959–1980.*
SAP (1965). L, June 5 (Silvano Arieti to Arthur Rosenthal). In *Subject File, 1914–1981. World Biennial of Psychiatry and Psychotherapy, editor. Correspondence.*
SAP (1966a). L, April 1 (William Dement to Silvano Arieti). In *Subject File, 1914–1981. American Handbook of Psychiatry, editor. Correspondence, 1956–1980.*
SAP (1966b). L, November 20 (Silvano Arieti to Konrad Lorenz). In *Subject File, 1914–1981. World Biennial of Psychiatry and Psychotherapy, editor. Correspondence.*
SAP (1966c). L, December 8 (Arthur Rosenthal to Silvano Arieti). In *Subject File, 1914–1981. World Biennial of Psychiatry and Psychotherapy, editor. Correspondence.*
SAP (1966d). L, December 13 (Konrad Lorenz to Silvano Arieti). In *Subject File, 1914–1981. World Biennial of Psychiatry and Psychotherapy, editor. Correspondence.*
SAP (1967a). L, February 7 (Giuseppe Moruzzi to Silvano Arieti). In *Subject File, 1914–1981. World Biennial of Psychiatry and Psychotherapy, editor. Correspondence.*
SAP (1967b). L, February 23 (William H. Masters to Silvano Arieti). In *Subject File, 1914–1981. World Biennial of Psychiatry and Psychotherapy, editor. Correspondence.*
SAP (1968a). L, April 18 (Ronald Laing to Silvano Arieti). In *Subject File, 1914–1981. World Biennial of Psychiatry and Psychotherapy, editor. Correspondence.*
SAP (1968b). L, August 21 (Arthur Rosenthal to Aleksandr R. Lurija). In *Subject File, 1914–1981. World Biennial of Psychiatry and Psychotherapy, editor. Correspondence.*
SAP (1968c). L, September 18 (Aleksandr R. Lurija to Arthur Rosenthal). In *Subject File, 1914–1981. World Biennial of Psychiatry and Psychotherapy, editor. Correspondence.*
SAP (1968d). L, December 28 (Silvano Arieti to Charles W. Whal).
SAP (1969a). L, August 16 (Eduardo Kalina to Silvano Arieti).
SAP (1969b). L, October 8 (Camille Laurin to Silvano Arieti).
SAP (1969c). L, November 5 (Gerard Chrzanowski to Alfred Freedman).
SAP (1969d). L, December 5 (Silvano Arieti to Julius Zellermayer). In *Subject File, 1914–1981. World Biennial of Psychiatry and Psychotherapy, editor. Correspondence, 1966–1970.*
SAP (1969e). L, December 24 (Silvano Arieti to Joseph Gabel).
SAP (1970a). L, January 13 (Jules Angst to Silvano Arieti). In *Subject File, 1914–1981. Conferences and Lectures, 1957–1981.*

SAP (1970b). L, April 21 (Silvano Arieti to Turan Itil). In *Correspondence, 1940–1981.*
SAP (1970c). L, June 19 (Silvano Arieti to David Hamburg). In *Subject File, 1914–1980. American Handbook of Psychiatry, editor. Correspondence, 1956–1980.*
SAP (1970d). L, December 29 (Silvano Arieti to The American Journal of Psychiatry; Psychiatry; Psychiatric Quarterly; The American Journal of Psychotherapy; The Journal of Nervous and Mental Disease). In *Subject File, 1914–1981. Organization. American Board of Psychiatry and Neurology.*
SAP (1970e). T (Silvano Arieti, *Committee Against Discrimination in Psychiatry*). In *Subject File, 1914–1981. Organization. American Board of Psychiatry and Neurology.*
SAP (1971a). L. January 28 (Joel M. Gora to Silvano Arieti). In *Correspondence, 1940–1981.*
SAP (1971b). L. February 9 (Silvano Arieti to Joel M. Gora). In *Correspondence, 1940–1981.*
SAP (1971c). L, March 7 (Silvano Arieti to Dieter Beck). In *Correspondence, 1940–1981.*
SAP (1971d). T, May 1 (Ian Alger, *The Origin and Effect of Power By Silvano Arieti, MD. Discussion By Ian Alger*). In *Speeches and Writings, 1940–1981. Lectures.*
SAP (1971e). L, June 13 (Silvano Arieti to Yrjo Alanen). In *Correspondence, 1940–1981.*
SAP (1971f). L, August 25 (Silvano Arieti to Walter Bruschi). In *Correspondence, 1940–1981.*
SAP (1971g). L, October 19 (Alonso Cantu Cantu to Silvano Arieti). In *Correspondence, 1940–1981.*
SAP (1971h). T (Silvano Arieti, *Revisiting the Concept of Transference and Counter-Transference in the Psychoanalytic Therapy of Schizophrenia*). In *Speeches and Writings, 1940–1981. Lectures.*
SAP (1972a). L, March 30 (Silvano Arieti to Jack Lynch). In *Correspondence, 1940–1981.*
SAP (1972b). L, April 17 (Gardner Spurgin to Silvano Arieti). In *Subject File, 1914–1981. World Biennial of Psychiatry and Psychotherapy, editor. Correspondence.*
SAP (1972c). L, May 18 (Simon H. Nagler to Luigi Boscolo). In *Correspondence, 1940–1981.*
SAP (1972d). L, May 31 (Silvano Arieti to Lucy Kroll).
SAP (1972e). L, June 15 (Silvano Arieti to Jack Lynch). In *Correspondence, 1940–1981.*
SAP (1972f). L, June 22 (Silvano Arieti to Jack Lynch). In *Correspondence, 1940–1981.*
SAP (1972g). L, July 17 (Alfonso Mangoni to Silvano Arieti).
SAP (1972h). L, July 22 (Silvano Arieti to Arthur Rosenthal). In *Correspondence, 1940–1981.*
SAP (1972i). L, August 2 (Silvano Arieti to Umberto Eco).
SAP (1972j). L, August 25 (Silvano Arieti to Erwin Glickes). In *Correspondence, 1940–1981.*
SAP (1972k). L, September 21 (Silvano Arieti to Simon H. Nagler). In *Correspondence, 1940–1981.*
SAP (1972l). L, October 4 (Francis Braceland to Silvano Arieti). In *Correspondence, 1940–1981.*
SAP (1972m). L, October 20 (Giorgi Giotto Immobiliare to Silvano Arieti). In *Correspondence, 1940–1981.*
SAP (1972n). L, November 1 (Silvano Arieti to Herbert Nagourney). In *Speeches and Writings, 1940–1981. Will to Be Human, Correspondence, 1971–1973.*
SAP (1972o). L, December 19 (Gaetano Benedetti to Silvano Arieti). In *Correspondence, 1940–1981.*
SAP (1972p). T (Silvano Arieti, *Silvano Arieti Introduces Dr. Keniston*). In *Speeches and Writings, 1940–1981. Lectures.*
SAP (1972q). T (Silvano Arieti, *Silvano Arieti Introduces Professor Benedetti*). In *Speeches and Writings, 1940–1981. Lectures.*
SAP (1973a). L, January 2 (Serge King to *Psychology Today*). In *Correspondence, 1940–1981.*
SAP (1973b). L, January 3 (Walter Bruschi to Silvano Arieti). In *Correspondence, 1940–1981.*
SAP (1973c). L, January 22 (Dargut Kemali to Silvano Arieti).
SAP (1973d). L, January 23 (Natalie Shainess to Elizabeth Hall).
SAP (1973e). L, March 8 (Silvano Arieti to Eugene Brody). In *Correspondence, 1940–1981.*
SAP (1973f). L, March 19 (Herbert Nagourney to Silvano Arieti). In *Speeches and Writings, 1940–1981. Will to Be Human, Correspondence, 1971–1973f.*
SAP (1973g). L, March 23 (Natalie Shainess to Silvano Arieti). In *Correspondence, 1940–1981.*
SAP (1973h). L, August 14 (Louis Joylon West to Silvano Arieti). In *Correspondence, 1940–1981.*
SAP (1973i). L, September 4 (Silvano Arieti to Louis Joylon West). In *Correspondence, 1940–1981.*

SAP (1973j). L, November 15 (Louis Joylon West to Silvano Arieti). In *Correspondence, 1940–1981.*

SAP (1974a). L, February 19 (Silvano Arieti to Louis Joylon West). In *Correspondence, 1940–1981.*

SAP (1974b). L, March 18 (Louis Joylon West to Silvano Arieti). In *Correspondence, 1940–1981.*

SAP (1974c). L, September 20 (Aldo Ghezzani to Silvano Arieti). In *Correspondence, 1940–1981.*

SAP (1974d). T (Silvano Arieti, *Psychiatric Controversial Attitudes Toward Man's Ethical Dimension*).

SAP (1975a). L, March 15 (Silvano Arieti to Erwin Glickes). In *Correspondence, 1940–1981.*

SAP (1975b). L, April 15 (Silvano Arieti to Arnold L. Wyse). In *Correspondence, 1940–1981.*

SAP (1975c). L, September 11 (Léon Chertok to Silvano Arieti). In *Subject File, 1914–1981. Conferences and Lectures, 1957–1981.*

SAP (1975d). L, December 23 (Silvano Arieti to David J. Halliday). In *Correspondence, 1940–1981.*

SAP (1976a). L, April 30 (Dargut Kemali to Silvano Arieti).

SAP (1976b). L, June 7 (Tina Zanotti to Silvano Arieti).

SAP (1976c). L, June 22 (Silvano Arieti to Evelyn Quinn). In *Correspondence, 1940–1981.*

SAP (1976d). L, October 25 (Benjamin Wolman to Silvano Arieti). In *Correspondence, 1940–1981.*

SAP (1976e). T (Università di Napoli, *Elenco dei docenti e relativi seminari per la Scuola di Specializzazione in Psichiatria*).

SAP (1976f). T (Università di Napoli, *Scuola di Specializzazione in Psichiatria. Seminari di aggiornamento 1976f–1977. Programma*).

SAP (1977a). L, March 9 (James H. Ryan to Silvano Arieti). In *Correspondence, 1940–1981.*

SAP (1977b). L, March 29 (Silvano Arieti to Therese Benedek). In *Speeches and Writings, 1940–1981. Love Can Be Found. Correspondence, 1976–1977b.*

SAP (1977c). L, May 10 (Carlo Gentili to Silvano Arieti). In *Correspondence, 1940–1981.*

SAP (1977d). L, June 7 (Marion Kift to Silvano Arieti). In *Correspondence, 1940–1981.*

SAP (1977e). L, August 6 (Joan Kirtland to Silvano Arieti). In *Correspondence, 1940–1981.*

SAP (1977f). L, September 23 (Jacky Korer to Silvano Arieti). In *Correspondence, 1940–1981.*

SAP (Ua). IC (Arnaldo Forlani to Silvano Arieti). In *Correspondence, 1940–1981.*

SAP (Ub). T, Unt. (Notes on the anti-psychiatric movement). In *Subject File, 1914–1981. American Handbook of Psychiatry, editor. Second edition. Vol. 1, circa 1972.*

Siirala, M. (1963). Schizophrenia: A Human Situation. *The American Journal of Psychanalysis,* 23 (1), 39-66.

Skinner, B. F. (1971). *Beyond Freedom and Dignity.* Knopf.

Sullivan, H. S. (1955). *Conceptions of Modern Psychiatry. The First William Alanson White Memorial Lectures.* Tavistock Publications.

Titchener, E. (1929). *Systematic Psychology: Prolegomena.* Mac Millan.

Trethowan, W. H. (1968). The American Handbook of Psychiatry. Edited by Silvano Arieti. *British Journal of Psychiatry,* 114 (506), 129–131.

Wolman, B. (Ed.). (1977). *International Encyclopedia of Psychiatry, Psychology, Psychoanalysis & Neurology.* Aesculapius.

Chapter 6
Landings

6.1 New Maps

Important changes occurred in American psychiatry in the 1970s. "Multiple ideological divisions" (Sabshin 1990, p. 1270) had started to sprout by the middle 1960s, especially between the biological-medical and the sociological approaches. As psychoanalysts were torn between orthodox and independent schools of thought, so psychiatrists went through infighting in their different schools.

In 1969, in his presidential message to the Society of Medical Psychoanalysts, Arieti addressed the discord and invited his colleagues to avoid both sociological extremism—which aimed at breaking the bonds with the medical tradition of psychiatry—and biological extremism, which rejected eclecticism for the sake of strictly organic explanations.

Pointing out the challenge that psychiatry was called to deal with—preserving its medical identity without yielding to biological reductionism, in his speech Arieti said:

> Some people may believe that eclecticism means giving up scientific standards. In fact there cannot be various point of view when scientific truth is found. Perhaps this way of thinking is correct in relation to many sciences, including many biological sciences. But certainly it cannot refer to a science to which the traditional scientific methods are not applicable. Objectivation and quantification which exclude ultimately any form of eclecticism cannot be applied to a science which deals with the private, the subjective, the felt experience and the unpredictability of human individuality. This eclecticism must continue to live and to grow. We must close our ranks (Arieti 1969, p. 30).

In 1972 Francis Braceland also was thinking about this challenge, confiding to Arieti his worries about the increasingly partisan approaches in American psychiatry. After having struggled to be recognized as physicians, psychiatrists were now divided into different groups. On one side were those who leaned toward neurology and who had started to talk about making a separate Board of their own; on the other, "the young man all gung ho about sociologic factors" (SAP 1972e). "I see a danger of some of us ending in the outer reaches of community and sociology ideologies, and

R. Passione, *Psychiatry and the Human Condition*, Springer Biographies,
https://doi.org/10.1007/978-3-031-09304-3_6

the other portion of us getting closer and closer to medicine" (SAP 1972e), Braceland observed.

In 1974, the second edition of *Interpretation of Schizophrenia* appeared in this environment (Arieti 1974a), and in it Arieti tried to counteract ideological polarizations by offering a wide-ranging study that reasserted integration as the key method of psychiatry (Balbuena Rivera 2016).

For the new edition Arieti returned to where he had started his exploration of schizophrenia, Pilgrim State Hospital, for in his daily medical practice he could not readily find so large a number of clinical cases as in the past, since New York Medical College's Psychoanalytic Unit did not admit schizophrenic patients (SAP 1972b); moreover, severely regressed patients were very rarely available in private practice. So, in the warm summer of 1968, accompanied by his wife Marianne, Arieti again entered Pilgrim State, where he was admitted as an observer by Henry Brill, director of the Hospital (SAP 1968a, 1968b, 1976h).

In these years, as has been discussed (Chaps. 4 and 5), Arieti worked on many projects simultaneously. He focused most of his attention, however, on this book. "I look forward to the end of this colossal work which continues to drain my energy," he wrote in September 1973 to Iris Topel, Editorial Supervisor at Basic Books; "I am afraid, however," he continued, "that as soon as I finish this work, my obsessive–compulsive neurosis will compel me to do something else" (SAP 1973c). In the meantime, even during summer, he visited a few Italian psychiatric hospitals to make clinical observations on severely regressed patients (SAP 1973d; Arieti 1974a).

The new edition of *Interpretation of Schizophrenia,* entirely rewritten and largely expanded, was published at the end of 1974. It was a major endeavor, incorporating the experience and knowledge he had gained in more than thirty years of work. The book was nominated for the National Book Award in Science. Supported by *The New Yorker* magazine, as well as by the physicist and science writer Jeremy Bernstein, the book was awarded the prize (SAPf; Fig. 6.1).

"The fact that it is the first time a book in psychiatry has received the award really calls for a celebration," Francis Braceland wrote (SAP 1975c). It was also "the first time that a book in psychiatry received the award *in the field of science*" (SAP 1975b).[1]

"A monumental and definitive study of what is known and conjectured about schizophrenia—remarkable both for its scientific content and profound humanism. No one who reads Dr. Arieti's long book will go away unmoved. The human mind is a strange and mysterious continent and Dr. Arieti has illuminated some of its darkest parts" (SAP 1975b). With these words the judges presented the award to Arieti at

[1] The other nominees in Science were: L.S. Feuer, *Einstein and the Generations of Science;* H. E. Gruber, P.H. Barret, *Darwin on Man: A Psychological Study of Scientific Creativity;* J.L. Heilbron, *H.G. J. Moseley*: *The Life and Letters of an English Physisicist, 1887–1915;* R.S. Lewis, *The Voyages of Apollo. The Exploration of the Moon;* J. McPhee, *The Curve of Binding Energy: A Journey Into the Awesome and Alarming World of Theodore B. Taylor;* S. Milgram, *Obedience to Authority: An Experimental View;* W. Sullivan, *Continents in Motion: the New Hearth Debate*; L. Thomas, *The Lives of a Cell: Notes of a Biology Watcher;* D.B. Vitaliano, *Legends of the Earth: Their Geologic Origins.*

Fig. 6.1 Silvano Arieti in his office with the National Book Award, 1975 (Personal Collection of James Arieti) © Courtesy of James Arieti—All rights reserved

a ceremony held at Lincoln Center, New York, on April 16, 1975, at 6 p.m. (SAP 1975a). It was a significant and unforgettable moment for Arieti, who accepted the prize before an audience of more than a thousand with "the joy, the trepidation, the rebirth experience of people whose work has been recognized." "This work of mine with schizophrenic patients," Arieti continued, "has been and continues to be not just part of my profession, but of my blood, my soul, the meaning of my life" (Arieti 1975a).

Significantly, in his speech Arieti connected his work to the theme of loneliness, which had been at the heart of his dialogue with Frieda Fromm-Reichmann (Chap. 3). Also in the *Preface* of the book he had dwelled upon the solitary nature of his endeavor in a contrast with his work on the *Handbook*:

> As I reflect on the difference between this work and the other one that has required a great deal of my time—the editorship of the *American Handbook of Psychiatry*—the following though emerges. ... The present book is the work of one man. Although I learned much from teachers, colleagues and other writers, I paved my own way down the various avenues of this vast subject. I am pleased that I did not find it necessary to seek financial support from either taxpayer money or foundation funds. Thus, for the errors, as well as for the new insights expressed in these pages, I alone must be held accountable. Be lenient, reader, but not too much; for I was not alone in this thirty-three-year work. Always with me was the sufferer (Arieti 1974a, p. viii).

Arieti's remark is suggestive; behind this "one man's" words lies the claim of a scientist who paved his own way. Also his reference to financial independence had an implicit lament for he actually had sought financial support, but without success (SAP 1969a, 1969b, 1969c).[2]

The solitary character of Arieti's venture was pointed out again in an article in *Time* magazine: "It is useful to be reminded that books that matter still tend to be created by individuals working over long periods of time—mostly alone and not mainly for money" (Kalem 1975, p. 95). This remark is reminiscent, by contrast, of Thomas Kuhn's concept of "normal science" (Kuhn 1962)—and Arieti's science was surely not "normal science."

As mentioned, *Interpretation of Schizophrenia* appeared in a milieu of an increasingly harsh confrontation between biological and sociological interpretations of mental illnesses. This ideological polarization was reflected in radically different approaches to schizophrenia: those who described it as the result of social factors and asserted its non-medical nature, and those who conceived it in strictly medical terms—as a biological, biochemical, or genetic phenomenon. As many reviews pointed out, Arieti's book represented an original voice, for it indicated a third possible path for psychiatry (McReynolds 1975; SAPg). In those pages he replied to both extreme positions. First, he criticized the excesses of strictly biological conceptions, which, according to him, had produced debatable results in even the most promising fields of genetics and biochemistry. In contrast, he emphasized the importance of those dynamic conceptions that had produced major advancements in the understanding and the treatment of schizophrenia: "It is impossible to overestimate the value of the dynamic approach in schizophrenia and, of course, in psychiatry in general. Nothing could be more important" (Arieti 1974a, p. 5), he wrote. This approach had made it possible to conceive symptoms not only as objective and natural expressions of the disease, but also as subjective phenomena related to an individual's history and life. Thus, in an era marked by the progress of biological conceptions and pharmacological therapy, with his book Arieti restated the importance of the dynamic approach and promoted further developments of psychotherapy by suggesting innovative treatments of schizophrenia—for example, the participation of a "therapeutic assistant" in the patient's daily life (Arieti 1974a; Arieti & Lorraine 1972; SAPj).

However, the belief that mental illness is exclusively a biological phenomenon was just one of the psychiatric ideologies of those years, and with the second edition of *Interpretation of Schizophrenia* Arieti replied also to psychiatrists questioning the very concept of mental illness. In particular, he discussed Thomas Szasz's view, which had become a pillar of the anti-psychiatric movement since the 1960s (Szasz 1957, 1961). Even before the publication of his famous *The Myth of Mental Illness* in 1961, Szasz had questioned the medical conception of mental illnesses by calling the word schizophrenia "a *panchreston*, a term coined by Hardin to denote dangerous

[2] In 1969 Arieti contacted the Center for Studies of Schizophrenia of the National Institute of Mental Health, directed by Loren Mosher, and the Foundations' Fund for Research in Psychiatry, directed by Roy Grinker, to ask financial support for his book. The Foudations' Fund replied as follows: "Dear Doctor Arieti ... I doubt whether our current Board of Directors would be interested in helping you with this. FFRP has traditionally been interested in more basic research—the collection of new empirical data .... I feel they would consider your project somewhat outside the scope of their interest".

words that are purported to explain everything, but that actually obscure matters" (Arieti 1974a, p. 692). As Arieti pointed out, according to Szasz, the problem of schizophrenia in psychiatry could be compared to that of ether in physics, since both were not "real facts" but artificial and hypothetical conceptions. Schizophrenia, Szasz maintained, was therefore just an empty word, used "to fill a scientific void" (Arieti 1974a, p. 692) and to label and stigmatize the existential condition of a person (Arieti 1960, 1971, 1974a).

In contrast, Arieti accepted that schizophrenia was a "real fact," an actual illness that differed *qualitatively* from a "normal" and healthy condition: "I believe that when psychiatrists examine typical cases, for example, a patient who states that he is Jesus because he drank Carnation milk and therefore has been reincarnated, or who uses peculiar neologisms … or typical word-salad, or who sees everywhere FBI agents spying on him, or hallucinates all the time, or is in catatonic postures, or complete withdrawal, they are confronted with a constellation or gestalt that cannot be confused" (Arieti 1974a, p. 693).

The same distinction between normal and pathological conditions is at the basis of Arieti's criticism of another leading figure of the anti-psychiatric movement, the Scottish psychiatrist Ronald Laing. Despite appreciating Laing's views about mental illness developed in 1960 (Laing 1960)[3]—so much so that he had invited him to write for *The World Biennial*—Arieti could not accept the "political" conception of schizophrenia Laing suggested in 1967 in *The Politics of Experience* (Laing 1967):

> In Laing's opinion schizophrenia is not a disease, but a broken-down relationship ... The environment of the patient is so bad that he has to invent special strategies in order "to live in this unlivable situation." The psychotic does not want to do any more denying. He unmasks himself; he unmasks the others. The psychosis thus appears as madness only to ordinary human beings ... Not only the family but society at large with its hypocrisies makes the situation unlivable. Echoing in a certain way Szasz (1961), Laing goes to the extent of saying that the diagnosis of schizophrenia is political, not medical (Arieti 1974a, p. 126).

Despite acknowledging, as Siirala had, the ethical value of the schizophrenic's existential condition, Arieti did not agree with Laing's idea that schizophrenia was just a "normal reaction to an abnormal situation" (Arieti 1974a, p. 126). In his book he wrote:

> "Contrary to Laing's conceptions, in by far the majority of cases we cannot consider the patient ... as an asserter of truth, a remover of the masks. The patient tells us his experiential truth, which often contains some truth about the evils of the world. This partial truth must be recognized by the therapist and must be acknowledged and used in treatment. Its import must be neither ignored nor exaggerated. If we ignore it, we become deaf to a profound message that the patient may try to convey. If we exaggerate it, we also do a disservice to him. We may admire the patient for removing the masks, for saying what other people do not dare to say, for how much he accepted and how much he rejected ... But we must also recognize that the fragments of truth he uncovers assume grotesque forms, ... so that whatever insight he has achieved will be less pronounced and less profound than his distortion. And his distortion not only has no adaptational value, but is inimical to any form of adaptation even within a liberal community of men (Arieti 1974a, pp. 127–128).

[3] In his analysis of schizophrenia, for example, Arieti referred to the concept of "ontological insecurity" suggested by Ronald Laing in *The Divided Self*.

Thus, Arieti's idea that schizophrenia would mirror the malaise inherent in the human condition was radically different from those who lost sight of the pathological nature of the schizophrenic's way of coping with the vicissitudes of life. Along the same lines, Arieti replied to Gregory Bateson, who, with his theory on "double-bind," had described schizophrenia as the result of a communicative short-circuit experienced in infancy and childhood (Bateson, Jackson, Haley & Weakland 1956):

> The mother tells the child, "Pull up your socks." At the same time, her gesture implies, "Don't be so obedient." In this situation the child receives the message, "Pull up your socks," but if he does so, he is too obedient. The other message says, "Don't be so obedient," but if he does not obey, he will incur mother's disapproval. In colloquial expression, he is damned if he does it, and he is damned if he does not. ... Bateson and associates state that double-bind situations provoke helplessness, fear, exasperation, and anxiety in the individual. According to them, the schizophrenic early in his life was exposed to a great many double-bind situations. The eventual psychosis may be viewed as a way of dealing with double-bind situations. ... My criticism ... is that every man, normal, neurotic, and future psychotic, was exposed in childhood and later to double-bind situations. ... We must emphasize that double-bind situations represent not necessarily pathology, but the complexity of human existence. If we were called upon to deal, not with double binds, but only with single messages ... life undoubtedly would be much simpler and would offer much less anxiety, but it would not be human life; it would be unidimensional life (Arieti 1974a, pp. 98–99).

Another original feature of Arieti's thought concerned the evaluation of the role of the family, that in the sociological conception of schizophrenia was traditionally described as the main soil of psychodynamic pathological mechanisms. Well beyond Laing's and Bateson's contributions, this subject had been widely debated within psychodynamic school, where the evidence of a close connection between schizophrenia and family issues had emerged for the first time. In this respect, far from denying the pathogenic role of family psychodynamics, Arieti claimed to have highlighted it even before Laing (Arieti 1973a). Nevertheless, in his book he also urged emending the radicalizations of this view, for he believed that the individual unceasingly and actively alters in his mind the experience of his family relationships. In 1975 Arieti wrote:

> The child does not just reflect or absorb from the environment; he also tries to select what to absorb or what to give prominence to. The image that the child has of himself does not consist of reflected appraisals from parents or family members but of what the child did with those appraisals. ... There is a definite discrepancy between the way reality and the significant people in one's life were in the past and the way one has perceived them ... This discrepancy if one manifestation of individuality (Arieti 1975b, p. 41).

Ascribing to Harold Searles the merit of having emphasized the positive aspects of the relationship between the schizophrenic and his mother (Searles 1958) when the usual focus was on the idea of "evil mother," in *Interpretation of Schizophrenia* Arieti revised the concept of the so-called "schizophrenogenic mother"—a concept he himself had once endorsed (Arieti 1974b, 1974c, 1977a). Therapists who believed what patients have told them about their mothers as malevolent creatures and "monstrous human beings," "have made a mistake reminiscent of the one made by Freud when he came to believe that neurotic patients had been assaulted sexually by their parents. Later Freud realized that what he had believed as true was, in by far the majority of cases, only the product of the fantasy of the patient" (Arieti 1974a, p. 82). In this respect, Philip May, professor of psychiatry at UCLA, wrote to Arieti: "I was particularly impressed … with your drawing attention to the serious error that people make in assuming that parents are indeed exactly as their children describe them. I hope that your words will have wide influence and induce a certain sense of humility. I wish I could think that they would have an effect on child psychiatrists who are so prone to boast or recommend *parentectomy*" (SAP 1977a).

The revision of the concept of "schizophrenogenic mother" was emblematic of Arieti's criticism of those socio-cultural approaches that focused only on external factors and neglected intrapsychic processes (SAP 1974),[4] the analysis of which could help to understand in a better and more balanced way the role played by social factors in mental illness. *Interpretation of Schizophrenia* can thus be considered as an attempt to re-establish psychiatry's medical identity by its investigation of internal factors other than genetic or biochemical processes.

This attempt also characterized Arieti's work on depression, to which in 1978 he devoted another book, written with his cousin Jules Bemporad (Arieti & Bemporad 1978).

Arieti had already been working on this subject for some time. In 1957, at the annual meeting of the American Psychiatric Association, he had presented his early observations on depression, the psychodynamics of which he described, to a certain extent, as the opposite of schizophrenia:

> It appears that the early infancy and childhood ... of the future manic-depressive patient, are not as traumatic as those of schizophrenics ... The child finds acceptance and care. The mother is duty-bound and administers a lot of care to the new born. The child seems receptive to the influence of the significant adult; there are no autistic tendencies or attempts to prevent socialization ..., but on the contrary there is an immediate acceptance of the symbolic emotional world of the surrounding adults. The "thou", that is the other, is immediately accepted by the child ... The parents expect a great deal from him. Their attitude toward life in general urges them to evoke in the child an early sense of duty, responsibility, guilt: what is to be obtained is to be deserved.
>
> Although the manic-depressive pattern is less devastating than the schizophrenic, it leads to self-defeating defenses. In many cases ... the seemingly self-imposed burdens confer a very unhappy feeling to the patient who sees no alternative. Moreover, the necessity to please others and to act in accordance with the expectations of the others, confers a feeling of futility and emptiness. In fact the patient ... does not listen to his own wishes, he does not know what it means to be himself; he feels frustrated, unhappy. But these feelings of unhappiness, futility and unfulfillment are misinterpreted again. The patient feels that he is to be blamed for them. If he is unhappy, if he finds no purpose in life, it must be his fault, or he must not be worthy of anything else. A vicious circle is thus established which repeats itself and increases in intensity, often throughout the life of the patient, unless fortunate circumstances of psychotherapy intervene (SAP 1957d).

[4] In 1974 the American Psychanalytic Association invited Silvano Arieti to participate, as a representative of the socio-cultural approach, in a series of "Psychoanalytic Dialogues." Arieti replied promptly: "Thank you very much for inviting me …. I feel honored and glad to accept. … However, I do not want to be considered the representative of a socio-cultural approach exclusively. You may know that I give great importance to intrapsychic factors also."

In his analysis Arieti also connected the psychodynamics of depression with statistical data concerning its geographical distribution; whereas schizophrenia was most often found in highly industrialized countries and big cities, depression seemed more frequent in countries characterized by a traditional culture and a mostly rural social system.[5] In order to explain the reason for this discrepancy—since it was important to investigate why these conditions occurred more or less frequently in different areas—Arieti relied on David Riesman's *The Lonely Crowd*, published in 1950 (Riesman, Glazer & Denney 1950).

As a sociologist, Riesman paid particular attention to the analysis of thought and cognition in different societies. Influenced by Erik Erikson's psychosocial approach (SAP 1957c), in his work Riesman focused not only on social macrosystems but also on the corresponding different types of personality. In 1950 he suggested a distinction between two different social structures, which he connected with two different types of personality—the "other-directed" society, characterized by a general anomie and by an uncertain and unstable family environment; and the "inner-directed" society, where family roles were more clear-cut and organized but also more prescriptive and authoritative in their mandatory rules. Riesman connected these different social structures with different types of personality.

Taking his cue from these observations (SAP 1956a, 1957a, 1957b), Arieti applied Riesman's analysis of the inner-directed personality to the study of the manic-depressive psychosis:

> Riesman explains the establishment of the inner-directed personality as the result of demographic and political changes. At certain times in history a rapid growth of population determines a diminution of material goods and a psychology of scarcity. Although this type of psychology has occurred several times in history, we are particularly concerned with the reoccurrence of it which had its beginning at the time of the Renaissance and of the Reformation. ... The religious doctrines of Luther and, indirectly, those of Calvin, gave the individual the feeling that everything depends on his own efforts. The concepts of responsibility, duty, guilt and punishment acquired tremendous significance and came to color every manifestations of life. This type of culture, thus, which originated during the Renaissance and developed during the Reformation, swept sooner or later all West countries, first the Protestants, then the Catholics, and only relatively recently is being replaced by another type of culture, the other-directed. Whereas in some countries, like the United States, this replacement is taking place at a rapid speed, in other countries, like Italy, it takes place more slowly.
>
> In the inner-directed society the parent is duty-bound, and therefore at the time of the birth of the child is very much concerned with his care and very devoted to him. ... Soon, however, this state of bliss will end for the child. The same duty-bound parent will start to burden the child with responsibilities, sense of duty, and guilt. This is often made necessary by the fact that the mother has now to devote herself to another child in these numerous families. The Paradise is lost; life must be a Purgatory to regain the Paradise. A personality develops which is inner-directed in the sense that it is implanted early in life by the direct and definite influence of the significant adults. As Riesman writes, this type of personality is directed toward "inescapably destined goals." It acts almost as if it would be equipped

[5] Arieti referred to Italian statistics, provided by his Italian colleague Francesco Bonfiglio, and to American statistics as well. In the U.S. the number of manic-depressive patients exceeded the number of schizophrenics only in rural States like Kansas, Tennessee, Alabama, and South Dakota.

> with a psychological gyroscope which is set up by the parents and keeps the inner-directed person "on course." ... The similarities between the pattern of life of the manic-depressive patient and the typical inner-directed person are apparent (SAP 1957d).

Lest he be misunderstood, however, Arieti pointed out that he did not mean that the inner-directed culture was the cause of manic-depressive psychosis. Far from saying that mental illnesses were the mere results of social factors, he was stressing that the type of culture analyzed by Riesman was "more liable to elicit a family configuration and interpersonal conflicts" that were "generally those which lead to manic-depressive psychoses"(SAP 1957d).

This paper, presented in 1957 to the American Psychiatric Association and published in 1959 (Arieti 1959), was the preliminary sketch of a broader study, written with Jules Bemporad, *Severe and Mild Depression: The Psychotherapeutic Approach*, published in 1978 (Arieti & Bemporad 1978).

Their co-authorship of this book resulted from a strong human and scientific relationship. In 1939 Silvano and Jules had arrived together in the United States; at that time, Jules was not yet two years old. Along with his brother Jack, he grown up in a close relationship with his elder cousin Silvano, who instilled in him a passion for the study of the human mind and thus introduced him to psychiatry.

Between the extremisms of the biological and sociological schools of thought, this book, like the second edition of *Interpretation of Schizophrenia*, offered a "third path" for psychiatry. On the one hand, it criticized the biological approach, that, owing to the psychopharmacological revolution sustained by Roland Kuhn's discoveries (Healy 1997, 2002; Ehrenberg 1998; Herzberg 2010), was dominant at that time in the treatment of patients. In contrast, Arieti and Bemporad stressed the need for a psychologically oriented approach to depression. Presenting the book before the public on the occasion of a promotional lecture, with these words Arieti referred to Nathan Kline's *From Sad To Glad* (Kline 1974):

> In these late 70s, when so much is said and written about depression, when the condition is generally treated with tricyclics and MAO inhibitors, when a prominent psychiatrist and a leader in the fight against depression states that a pill is the most effective method from changing the human being from sad to glad, my voice is one among the few to reaffirm the importance of psychological factors, both in the etiology and the therapy of this common condition (SAP Ue).

On the other hand, Arieti and Bemporad rejected the idea that only external social and environmental causes were at the root of depression. They rejected, for example, the concept of "reactive depression," generally used "to describe some forms of depression in which the precipitating event seems of major importance" (Arieti & Bemporad 1978, p. 6). According to them, an environmental influence assumes pathogenetic power only when the patient attributes a special meaning to it:

> This study of the depressed person will show how we ourselves can contribute to our own sorrow with the strange ways in which we mix and give meaning to our ideas and feelings. We shall learn that the study of life circumstances is important, but that even more important is the study of our ideas about these circumstances, our ideals and what we do with them, and how we use them to create feelings (Arieti & Bemporad 1978, p. 10).

In another paper on the subject, Arieti also talked about the essential role played by a "pre-existing ideology, a certain way to see oneself in life" (Arieti 1980a, p. 21, my translation) in bringing on depression. For the onset of this disease, therefore, a cognitive predisposition is needed, for the depressed person thinks in a way that molds his world—a world where external circumstances are filtered and transformed by pre-existing cognitive structures that originated from infancy and that continued to grow for the rest of his life.

Thus, turning to the sociology of knowledge—the branch of sociology that studied the relation between thought and society, Arieti and Bemporad suggested a synthesis of cognitive and sociocultural approaches:

> A cognitive approach lends itself better than any other to the integration of psychiatry and sociocultural studies. A cognitive approach stresses the importance of ideas, and most of a person's ideas derive from the sociocultural environment. Although a large number of thoughts and habits of thinking are required in childhood from the members of one's family, the family members are carriers of the culture to which they belong (Arieti & Bemporad 1978, pp. 361–362).

The importance of the cognitive factors that had been the focus of Arieti's thinking on schizophrenia was thus reaffirmed for depression. Despite their clinical and psychodynamic difference, both illnesses brought to the surface some basic structures of the human psyche, the complexity of which, according to Arieti, has no equal in the world or in the realm of science. This complexity, neglected by biased views, could be fully acknowledged by a psychiatry conceived as both a medical and a human science—that is, a psychiatry capable of going beyond a theoretical one-sidedness by acknowledging the many facets of human nature.

The second edition of *Interpretation of Schizophrenia* and *Severe and Mild depression* were Arieti's attempts to provide new maps for traveling through "the strange and mysterious continent" (SAP 1975b) of the human mind. With these books, Arieti aimed at re-establishing the medical identity of psychiatry without dismissing its distinctiveness as a discipline concerning the human condition. To understand this intention, two observations about the books that Arieti made are illustrative. The first he expounded at the National Book Award Ceremony:

> This award is much more than an appreciation of my work. ... In choosing this book the judges ... must have accepted my message that it is possible for the therapist to help the patient to experience the human tie more intensely than any fear, more strongly than any need for distance. ... We therapists are able to learn and practice ways of bringing trust to the distrustful, clarity to the bewildered, speech to the mute ..., confident expectation to the hopeless, and companionship to the lonely. And in so doing we can suggest a larger vista for the human horizon, a larger use for the human bond, and optimism for the solutions of those other conditions in health or illness that, although difficult, are not so obscure, so hard to approach, so desperate, so far from the usual reach of man's words and care (Arieti 1975a, p. 234).

The second appeared at the end of the first chapter of the book on depression:

> There is always a resonance in our heart for the anguish of the depressed which does not seem to us completely unfounded, but similar to ours, and containing a partial truth based on the human predicament. ... This study will explore, and we hope to some extent enlighten,

> not just our pathology but our so-called normality; not just our despair, but our confident expectation; not just our loneliness but also our ways of helping each other and reinforcing the human bond (Arieti & Bemporad 1978, p. 10).

These observations reasserted psychiatry as a discipline capable of going beyond narrow views to accomplish its therapeutic duties and become a discipline that contributed to the care and unity of mankind—a conception shared by a minority of psychiatrists in those years (SAP Ud). And yet, for this very reason, it was important that this voice be heard.

6.2 Compasses

With the second edition of *Interpretation of Schizophrenia* and with *Severe and Mild Depression,* Silvano Arieti contributed two significant works to the cause of a psychiatry capable of not sacrificing its epistemological distinctiveness on the altar of diverse exclusive principles, principles in themselves and by themselves insufficient to portray the complexity of human nature. With this cause in mind, Arieti regarded schizophrenia and depression not only as clinical issues but also as scientific and epistemological challenges for the future of psychiatry. He did not intend to break the bond between psychiatry and medicine, but he aimed at reasserting the medical identity of psychiatry by means of a renewal of the medical model (SAP 1977e)[6]—a renewal capable of rejecting medicine's growing biological and genetic reductionism and of taking into account the "biopsychosocial complexities of the patient's life situation" (SAP 1977b), an aim that calls to mind Adolf Meyer's early "holistic" teachings (Lidz 1966; Lamb 2014).

In encouraging psychiatry to preserve its medical identity without dismissing its own epistemological peculiarity, Arieti referred in *Interpretation of Schizophrenia* to one of the founding fathers of modern medical science—Rudolf Virchow. In particular, Arieti was questioning the application of cellular pathology to mental illnesses, in those years an important feature of psychiatry's biological reductionism:

> Can we state that schizophrenia is an illness? If we follow the concepts of Virchow, or those derived from Virchow, which imply cellular pathology, an understanding of the pathological mechanisms, and the capacity to reproduce experimentally the condition, the answer is no. ... Schizophrenia, as well as most mental illnesses or psychiatric conditions, does not fit the medical (especially Virchowian) model. This realization does not necessarily lead to the conclusion that the concept of schizophrenia or any mental illness is a myth. An alternative position is that the traditional medical model was built without taking psychiatry into consideration and does not include all the dimensions of human pathology. If we do change the traditional medical model, we can then call schizophrenia an illness (Arieti 1974a, p. 4).

So, the stakes were high and the challenge ticklish—they were about changing the traditional medical model without slipping, so to speak, out of medicine. In this

---

[6] In 1977 and 1978 The Forest Hospital Foundation promoted a series of scientific lectures on the renewal of the medical model, and Arieti was invited as guest of honor.

regard, the contribution of Claude Bernard—author of the classic *An Introduction to the Study of Experimental Medicine*, published in France in 1865 (Bernard 1865), and another founding father of medicine—could help. Arieti wrote about Bernard and his work:

> Since the work of Claude Bernard the usefulness or adaptational value of a pathological mechanism was recognized not only in psychiatric conditions, but in the whole field of medicine. In infectious disease, for instance, fever occurs as a reaction to the invasion of foreign proteins. This reaction can be interpreted in accordance with deterministic causality. Fever, however, seems to have a purpose: to combat the invasion of foreign proteins. Here the organism seems to follow a purpose, or teleologic causality. Only organisms that are able to build up adequate defenses can survive and transmit such a possibility genetically. Thus the defenses, from a human point of view, do acquire a purpose (Arieti 1974a, p. 223).

Arieti's reference to Claude Bernard had a double function: first, it aimed at restoring to medicine the relationship between diseases and adaptation—a theme that in those years was at the center of sociological explanations of mental illnesses[7]; second, it showed that teleology—an idea particularly criticized in the reductionism of biological psychiatry—was actually an inherent characteristic of living matter, one pertaining not only to psychological phenomena but also to the biological ones (Arieti 1974a; SAP 1970). This reasoning also drew the attention of Beppino Disertori, an Italian psychiatrist and philosopher, who saw in the acceptance of biological finalism a crucial step in the struggle against the reductionism that traditionally informed psychiatry and its scientific study of human beings (Disertori 1947; SAP 1967).

The medical model in psychiatry that Arieti intended to criticize, however, was not limited to the reductionist biological approach. In the very years that he was writing the second edition of *Interpretation of Schizophrenia*, in fact, the American Psychiatric Association was discussing the need for a new classification of mental diseases—a medical nosology aimed at standardizing the diagnostic criteria employed in clinical practice (Fischer 2012; Decker 2013; Shorter 2015). Its model was the *Diagnostic and Statistical Manual of Mental Disorders* (*DSM*), the first edition of which appeared in 1952, the second in 1968 (APA 1952, 1968). This second edition was discussed in 1969 at the Conference on the Schizophrenic Syndrome organized by the Menninger Clinic in Topeka, Kansas, attended by more than seven hundred psychiatrists coming from all over the country—Arieti among them (World Wide Medical News Service 1969; Fig. 6.2).

On that occasion, Robert Cancro, Director of Research Training at the Menninger Foundation, had spoken about the new *DSM* as a serious loss for American psychiatry. In particular, Cancro asserted, echoing Karl Menninger's adverse opinion of nosology (Menninger, Mayman & Pruyser 1963), that classifications could cause a barren petrification of psychiatry instead of a fecund theoretical development (World Wide Medical News Service 1969).

[7] In the 1960s and the 1970s the concept of adaptation was interpreted in different ways within the anti-psychiatric movement; Ronald Laing, for example, believed that schizophrenia was a normal adaptation to an abnormal situation, whereas other authors stated that adaptation was a diagnostic criterion, because, according to them, psychiatrists used to diagnose maladjusted person as mentally ill.

Fig. 6.2 Silvano Arieti at the Conference on the Schizophrenic Syndrome, Topeka, Kansas, 1969 (Previously published in *Roche Report*. Reprinted with permission from Roche)

For his part, Silvano Arieti did not share the same aversion to diagnostic categories, the practical value of which he acknowledged in medical reasoning—a cast of reasoning he was trying to protect from those who denied the very existence of mental illnesses (Arieti 1971). Nevertheless, he did agree with Cancro that rigid classifications could cripple clinical work, encouraging, if not facilitating, a superficial and static analysis in which "the patient is always seen in cross section" (Arieti 1974a, p. 12), where symptoms are seen as crystalized features of the disorder, and, as a result, the changing progress of an illness may be overlooked.

According to Arieti, nosology was a controversial area in which psychiatry seemed to "try desperately to make consistent what is inconsistent" (Arieti 1968, p. 1637). Given the lack of certain knowledge about the origins of most mental illnesses, every classification had necessarily to be based on a cluster of symptoms that segmented the pathological process into speciously simple and consistent phenomena. In 1968—the same year of the publication of *DSM*'s second edition, Arieti wrote:

> Another field which requires attention is that of psychiatric nosology and/or nomenclature. We are still following a classification of mental diseases which is more or less based on the original one by Kraepelin. The truth is that at the present stage of our knowledge we cannot do much better. Perhaps this is the only field in which further work should be postponed until greater knowledge of mental disorders is achieved. For the time being, we must do the best we can with the unsatisfactory classifications that are at our disposal (Arieti 1968, p. 1637).

The march to an even more standardized diagnostic system did not, however, come to a halt, largely because of the pressure exerted by insurance companies, which in the 1970s more and more insistently demanded a system capable of guaranteeing indisputable objectivity in clinical judgements and diagnoses. A first contribution towards this goal was the publication in 1972 of the "Feighner criteria" in *The Archive of General Psychiatry* (Feighner, Robins, Guze, Woodruff, Winokur & Munoz 1972; Kendler, Munoz & Murphy 2010), which led the American Psychiatric Association in 1973 to set up a task force to prepare a revised classification of mental diseases (the so-called *DSM*-III). Robert Spitzer of the New York Psychiatric Institute, where he also directed the "Biometric Research Unit," was nominated to head the project (Kirk & Kutchins 1992; Shorter 1997).

The job of the task force was to anchor diagnostic categories in descriptive, not theoretical criteria. To accomplish the job the task force needed to revise the previous classifications, which had been drawn up under the influence of dynamic psychiatry. Not coincidentally, most members of Spitzer's working group were not advocates of psychoanalysis, but were aligned with Washington University in St. Louis, an internationally known epicenter of the biological approach in psychiatry (Shorter 1997; Clayton 2006; Campbell 2014).

On June 11, 1976, the members of the task force gathered in St. Louis for the first "*DSM* in Midstream Conference," attended also by many other psychiatrists from all over the country (SAP 1976j). Arieti participated, a part of the skeptical audience that also included Howard Berk, who openly criticized the work of his "neo-Kraepelinian" (Sartorius 1990; Decker 2007) colleagues on the new *DSM*, maintaining that the *DSM*'s open rejection of etiological reasoning actually arose from their anti-psychological and anti-dynamic presuppositions. According to Berk, a useful diagnostic system already existed—the international classification of diseases (ICD-9) approved by the World Health Organization in 1975. The fifth chapter of the ICD-9 covered mental disorders and included a glossary that richly described the contents of diagnostic criteria (SAP 1975d; Decker 2013).

The glossary of the ICD-9 was useful in clinical practice, for it enhanced clinical judgement by requiring a careful evaluation of every single case. At the meeting in St. Louis it became a landmark for those who were critical of the new *DSM*:

> The ICD-9 with its more numerous opportunities for diagnosis and more varied classification ... may seem to those trying to create a structure of exquisite simplicity to be a farmer's market of diverse ideas. When one is trying to bring about a simpler order, a mélange of Reactions, Disturbances, Symptoms Syndromes, Dependences, States, Transient Reactions, Disorders, Psychosis, etc., may seem abysmal—and "unscientific," but ... it reflects the greater reality of the diversity that does exist in life and in psychiatry (SAP 1976k).

In particular, the difference between the ICD-9 and the *DSM* mirrored the difference between the concepts of nomenclature and classification, the latter less descriptive than the former. Concerning this distinction, Berk declared:

> Our differences with the Task Force's process and product become apparent when we examine a description of nomenclature and statistical classification and their crucial differences ... Nomenclature is very varied ... and seeks specificity in its effort to accurately represent the instance before it. It attempts to be a mirror of nature, simple or complex, in keeping with the observations ... Statistical Classification seeks to categorize, to generalize, to seek a common essence in the items of a nomenclature, and this in a way that reflects the belief, hypotheses, and goals of the designer of a particular statistical classification. ... Nomenclature, seeking identity with the object or phenomenon, is the more primary and

basic, and Statistical Classification is derivative, secondary and subordinate to Nomenclature (SAP 1976k).

The main shortcomings of the task force's endeavor was therefore the attempt to replace nomenclature with classification; in this way, according to Berk, Spitzer's work was like that of a director of a national museum of art who destroys Rembrandts, Goyas, and van Goghs, in order to provide room for Warhol's paintings of soup cans. Some might applaud so bold a move, but others would consider it "a tragic act of vandalism" (SAP 1976k).

After the Midstream Conference, Berk wrote to Arieti to ask him to exert his scientific reputation to counterbalance DSM's venture (SAP 1976c). At the Conference, Arieti had in fact expressed criticism of the work undertaken by the task force, showing an independent line of thinking. In particular, he was concerned about the task force's erratic use of words and concepts, for it seemed to him to mirror a diagnostic "behavioral attitude." A clear example was the limitation of psychotic disorders to delusional and hallucinatory states:

This ... reveals a gross misunderstanding of what psychosis is. It is true that a schizophrenic psychosis is often characterized by delusions and hallucinations, but a manic-depressive psychosis only seldom presents delusions and hallucinations, and still is a psychosis. The term psychosis indicates that a grossly psychopathological way of living is accepted by the patient. No matter what transformation the psychotic patient has undergone, that transformation becomes his way of relating to himself and others and of interpreting the world. The patient suffering from a psychosis does not fight his disorder, as does the psychoneurotic, but lives in it (SAP 1976l).

According to Arieti, the task force's focus on delusions and hallucinations revealed a bias toward overt behavior to the detriment of a deeper analysis of inner processes at the root of mental illness. This bias was also revealed in the replacement of the expression "affective disorders" with "mood disorders"—an exclusion from the diagnostic vocabulary of any reference to the affective dimension that mirrored the anti-dynamic turn of the new *DSM*. So, despite *DSM*'s purported agnosticism about etiological thinking, the terms the task force choose referred to given constellation of conceptions—a subject about which Arieti aimed to resume the debate.

Another of Arieti's criticisms of the classification was that the "simple type" of schizophrenia had been omitted:

I am disturbed about the omission of the "simple type." Althòugh rare, it does exist. I have seen several cases and so have the great masters, like Kraepelin, Bleuler, and Meyer. Elimination of the term is recommended by the Task Force because it implies a "nonpsychotic" form of the disorder. Again it depends on what we mean with the term psychotic. Psychotic is not exclusively the person who is delusional or hallucinated. The simple schizophrenic, although not delusional, shows a marked impairment of abstract thinking, reduces his life to a concrete life, and adopts this reduction ... as a normal way of living. Although he is not *obviously* psychotic, his life is limited and impoverished to a psychotic degree (SAP 1976l).

The case of "simple schizophrenia" showed better than any other example the principal shortcomings of the new *DSM*. It sacrificed a wealth of clinical experience for a uniform nosology that excluded those grey areas that allowed psychiatry to go beyond plain pathology. In short, the design of the new *DSM* consistently narrowed the reach of psychiatry—a limitation Arieti could not accept, for he firmly believed that "a psychiatrist cannot help but venture beyond psychiatry" (Arieti 1975c, p. 1, my translation).

Undoubtedly, the *Diagnostic and Statistical Manual of Mental Disorders* reflected the deep change occurring in American psychiatry—a change that concerned the very method of psychiatric reasoning (Wilson 1993). The principal result of the new *DSM* was not a simplification of diagnostic procedures by a reduction of diagnostic categories, the number of which was actually increased; rather, it resulted in a standardization of diagnosis *beyond* any clinical deliberation and individual expertise. Providing a common frame of reference for the complicated diagnostic process may have been an important task, but many critics believed that this endeavor had been undertaken simplistically, as Arieti said, by trying to make consistent what was inconsistent by nature. This kind of forced classification was not a proper compass to explore the troubled territory of madness.

In the 1970s Silvano Arieti promoted an alternative method of psychiatric reasoning with the founding of a journal dedicated to an understanding of human nature that was "not experimental and barren, and yet not pompous, apodictic and impermeable to wider horizons" (SAP Uc). In 1967, Arieti had suggested to Harold Lief, President of the American Academy of Psychoanalysis, that the Academy's series "Science and Psychoanalysis"[8] be replaced by a quarterly journal that would be a forum for scientific dialogues between different disciplines and approaches. This proposal came into being four years later, in 1971, when the Academy, under the directorship of Irving Bieber, approved the launching of a new journal with Arieti as Editor-in-Chief (SAP 1971a, 1971b, 1971c).

Sustained by a solid international Editorial Board,[9] Arieti set high standards of excellence in every issue (SAP 1972a, 1972c)—a goal that was recognized in 1973, when all the contributions appearing in the journal were selected for inclusion by the *Digest of Neurology and Psychiatry*. "This is an event not rare, but unique in editorial history" (SAP 1973b), Arieti remarked with great satisfaction.

Thus, in the same year that the task force started work on the new *DSM*, Arieti launched the journal, undertaking a challenge different from simple classification. Two different lines of development of American psychiatry confronted each other: one, a simplifying and standardizing approach; the other, an approach dealing with human complexity and engaging in dialogue (Arieti 1973b).

With *The Journal of the American Academy of Psychoanalysis* Arieti aimed at promoting a review of the scientific foundations of psychiatry and its relationship with the medical model. The contributions appearing in the journal between 1973 and 1981 (that is, under his directorship) reflected this attempt. In an article published in the first issue of the journal, for example, Harry Guntrip (Guntrip 1973; SAP 1971d,

[8] The series was edited by Jules Masserman and the first volume had appeared in 1958.

[9] The Editorial Board included Gaetano Benedetti (Switzerland), Annemarie Dührssen, Franz Heigl, Fritz Riemann (Germany), Emilio Servadio (Italy), and Martti O. Siirala (Finland).

1971e, 1972d, 1973a) took into account a provocative question raised by Eliot Slater in 1972—"Is Psychiatry a Science? Does It Want To Be?" Slater wrote:

> It is surely only the glamour the name of "science" exerts which has induced us all to mistake our functions. ... The scientific method can only concern itself with the real world, the world outside us, which we can to some extent study objectively. There is also the world within us, for ever the domain of subjectivity, for ever beyond the reach of science. (Slater 1972, p. 81)

Noting that "in the relativity atmosphere of Post-Einstein thinking … the ultimate reality of matter has become as mysterious as that of mind" (Guntrip 1973, p. 5), Guntrip replied to Slater that the "domain of subjectivity," far from being "beyond the reach of science," had instead to be studied with a wide-ranging integrative approach by entering "the experimental laboratory of the whole of human living" (Guntrip 1973, p. 21). Thus, psychiatry, to assert its place in medicine and science, had to reject reductionism and claim that the special complexity of human beings was its subject of study—a complexity that no "taxonomic stew" could abolish (Wolberg 1979). It was a challenge for psychiatry, and for science generally (Guntrip 1973; Redlich 1974; Rifkin 1974; Edel 1975; Eckardt 1977).

As the job of Editor-in-Chief of the journal markedly increased his workload, in 1976 Arieti asked Simon Nagler, director of training at the Department of Psychiatry of New York Medical College, to be relieved of supervising residents; "I have only one life to give to psychiatry and psychoanalysis," he wrote, paraphrasing American patriot Nathan Hale (SAP 1976i).

In truth, Arieti was experiencing a growing disaffection toward his teaching. Year after year, he felt the attitude of students was changing for the worse. They were the canary in a coal mine for the future of American psychiatry: they read less and less, were increasingly superficial, hasty, less attentive, and less curious. They seemed unable to concentrate, and in the classroom there was a constant coming and going of residents, to the detriment of any focused attentiveness. Arieti was distressed. "This state of deterioration is shocking to me," he declared; "I wonder whether the criteria for selecting residents should be changed or not" (SAP 1978b). He felt demoralized every time he went to teach, and he thought about quitting. Continuing to teach, he felt, would be "an exercise of masochism" (SAP 1978b).

At this point in his career, it was not from teaching that Arieti derived satisfaction. His satisfaction came instead from writing books, as well as from his service to the American Academy of Psychoanalysis and editing its journal, where he felt he could better contribute to the causes he believed (SAP 1977d).

In 1979 Silvano Arieti was elected President of the American Academy of Psychoanalysis. When his new position required that he suspend his job as editor of the *Journal*, Morton Cantor assumed the editorship (SAP 1978a, 1979b).

Arieti's presidency coincided with a turbulent time. As soon as he was elected there arose problems with the American Psychiatric Association and within the Academy itself.

The problem with the American Psychiatric Association concerned its increasing marginalization of psychological and dynamic approaches. At the end of the 1970s, for example, the new editorial guidelines of *The American Journal of Psychiatry* included reducing the length of articles and rejection of any submission that did not conform to this reduction. Arieti was very concerned about this change, which he felt would negatively affect articles' content. In 1979 he wrote to the Editor, John Nemiah, to express his objections to the new guideline:

> Dear Dr. Nemiah ... I wish to express my point of view that these space limitations almost automatically exclude most papers dealing with psychotherapy ... On the other hand, biological papers, dealing with drugs or with statistical data which can be tabulated in a few numbers, will have a great advantage. *The American Journal of Psychiatry* thus will represent more and more the biologically and pharmacologically oriented part of our field and will leave us out, psychotherapists, psychodynamic psychiatrists, and psychoanalysts. But we are psychiatrists, too. The classic papers by Freud, Jung, Bleuler, Adolph Meyer, Sullivan, and many other authors would not qualify for the journal if these space limitations were imposed on them. Perhaps many psychotherapists will be discouraged ... and will not send their papers to the Journal; or perhaps ... papers in this field will be too long to be considered. The result is that *The American Journal of Psychiatry* will be in danger of not representing any more the double nature of psychiatry, biological and psychological, which makes our field unique and so inspiring. ... How can we describe psychotherapy of schizophrenia in 15 double-spaced typewritten pages? I feel that people like me, members of the APA who are psychotherapists or psychodynamically oriented, should not be forced to beg hospitality elsewhere (SAP 1979m).

We are psychiatrists, too. With these words Arieti claimed the right of the dynamic perspective to be part of a discipline that was steadily moving more and more toward other approaches. In this atmosphere, of course, it was difficult to avoid creating new schisms while maintaining dialogue.

There was also much turbulence within the Academy itself. In those same years, a large number of members proposed admitting individuals who did not hold the degree of M.D. to the Academy, while another large part believed that the medical character of the organization should be maintained (SAP 1956b).[10] Arieti worried that allowing memberships to non-medical graduates would be interpreted as a move to detach psychiatry from medicine, and that, as a consequence, the expansion of membership might compromise his attempt to renew and reinvigorate the medical model for the field.

With these considerations, Arieti prudently decided to test the waters with past and present presidents of the American Psychiatric Association (SAP 1979c, 1979d, 1979e, 1979f, 1979g, 1979h). He wrote to Alan Stone, its current President, as follows:

> Dear Dr. Stone ... the Academy is going through a serious crisis just as I become president and it is incumbent upon me to do what I can to avoid a potential rift. Thus, this letter must be considered as an urgent appeal. ... Because of your relationship with the American Psychiatric Association, I feel you are in a particularly favorable position to answer the following questions: 1) How do you think the American Psychiatric Association would view our admission of psychologists and possibly other Ph.D.'s to our membership? 2) Do you feel such move would affect our relationship with the American Psychiatric Association—relations that, at this point in our history, we are trying to strengthen? 3) Do you

[10] The by-laws of the Academy granted admission of non-medical graduates only as "Scientific Associates" and not as members.

> believe that such move could weaken the position of the Academy to confer a certification in psychoanalysis recognized by the American Psychiatric Association and the American Medical Association? (SAP 1979c).

Stone replied politely, "Let me take no position on that" (SAP 1979i). The replies received from the APA's past presidents were more to the point: granting membership to non-medical graduates would be a professional and tactical mistake, as it would be viewed as a move of the Academy away from medicine (SAP 1979j, 1979k, 1979l).

In September, 1979, Arieti invited the membership to think about the proposal cautiously, for, if the amendment were approved, the Academy would change its status from a psychiatric association to a pluri-professional one, with many possible consequences (SAP 1979n).

Arieti had found himself in a similar situation in 1956, when a proposal had been made at the William Alanson White Institute to discontinue its training of psychologists and limit membership to graduates from medical schools. As he had in 1956 (Chap. 3), in 1979 he again supported the "medical party." As president of the American Academy of Psychoanalysis, he intended to do everything he could to strengthen the medical identity of psychiatrists. In short, he reaffirmed that psychiatry was a part of medicine—a special part, in which the study of the human being, when properly pursued, went beyond simply biological and statistical information.

6.3 Zenith

Silvano Arieti's first article in *The Journal of the American Academy of Psychoanalysis* was "Schizophrenic Art and Its Relationship to Modern Art" (Arieti 1973c).

Arieti's long interest in art acquired increasing importance for him in the 1970s, joining the range of topics that were part of his scientific inquiry. His attention to art was closely linked to his interest in cognition since early in his career. While he was at Pilgrim State Hospital, art itself prompted his study of cognitive functions when he realized that the art of schizophrenic patients was the expression of their peculiar way of thinking. "Their art work showed the unfolding of their illness toward more and more severe forms of disorganization. I have not forgotten these few cases … and I have kept their art works as precious possessions for many years, hoping the time would come when I would understand them better. That time, I believe, has come" (Arieti 1973c, p. 334).

The subject appeared in the first edition of *Interpretation of Schizophrenia,* where Arieti published a few drawings by schizophrenic patients. He described these drawings as a figurative expression of the word-salad typical of their illness: "When the schizophrenic produces artistic works, his lack of integration is obvious … The separate parts are put together in a sort of visual word-salad" (Arieti 1955, p. 313).

As is clear from his correspondence with Rudolph Arnheim, a German psychologist and art critic educated in Max Wertheimer's Gestalt psychology and author of several studies on the relationship of Gestalt theory and artistic form, Arieti continued to explore art and creativity after 1955, when *Interpretation of Schizophrenia* was published (SAP 1961a, 1961b, 1961c). At first, he had considered the idea of devoting his next book to this subject, but his many commitments, including the editorship of the *Handbook*, prevented him from carrying it through (SAP 1962).

In 1964, upon becoming president of the William Alanson White Society, Arieti addressed the study of art and creativity in his inaugural speech (SAP 1964; Arieti 1964). Shortly afterwards, he wrote a chapter on creativity in the third volume of the *Handbook* (Arieti 1966). Then, in 1967, he devoted the third part of *The Intrapsychic Self* to creativity in art and science (Arieti 1967). The fulfillment of these efforts took shape in 1976 in *Creativity: The Magic Synthesis* (Arieti 1976).

To collect as much empirical data for the book as possible, Arieti decided to return to Pilgrim State. Since it had been at the mental hospital that he had conceptualized both schizophrenia and art as attempts to change reality, he was returning to the scene of his interest in art and creativity (SAP 1977c). In 1980 he wrote:

> A combination of chance events and of my own choices has made me become intensely involved with two groups of people who confront their basic needs, aspirations, and troubles ... by attempting to transform reality. These two groups, the schizophrenics and the creative persons, are both fugitives from the daily reality in which they feel prisoners. The ones and the others are shaken by what they feel is terribly absent in this world, and they send us messages of their own search, and samples of their findings, that is, of the transformations which they themselves have put into effect. But ... the transformations of these two groups of people are quite different. The creative person wants to change reality in order to beautify it, or to enlarge the field of human knowledge or experience ... The schizophrenic instead is terribly afraid of this planet, ... he has no wings to fly into high space (Arieti 1980b, p. 287).

Schizophrenia and creativity therefore represent opposite poles, the nadir and the zenith of a common human condition that psychiatry could explore in its different facets. "The psychiatrist may become involved with two unusual and divergent transformation of reality: psychosis and creativity. If we pursue this area of inquiry in all its ramifications, we become astonished at the vast panorama of human existence which becomes available to the psychiatrist. He can look at certain forms of schizophrenia and touch the nadir of human existence, and may explore the harmonized fantasy of the creative person and participate in the zenith of human life" (SAP Ua).

The book, completed shortly after the printing of *Interpretation of Schizophrenia*'s second edition, was published by Basic Books in 1976. In it Arieti describes creativity as a means by which "the human being liberates himself from the fetters not only of his conditioned responses, but also of his usual choices" (Arieti 1976, p. 4), enlarging the universe "by adding or uncovering new dimensions" (Arieti 1976, p. 5). Thus, showing the limits of Skinner's deterministic view of human life, Arieti described how creativity involves freedom and originality. Still, it would be a mistake to summarize by these two words, "freedom" and "originality," the complicated nature and function of creativity:

> Creativity is not simply originality and unlimited freedom. There is much more to it than that. Creativity also imposes restrictions. While it uses methods other than those or ordinary thinking, it must not be in disagreement with ordinary thinking—or rather, it must be

something that, sooner or later, ordinary thinking will understand, accept, and appreciate. Otherwise the result would be bizarre, not creative (Arieti 1976, p. 4).

Here we find a clear distinction between schizophrenic art and creativity, since ordinary thinking, which Arieti described as an essential ingredient of artistic work, is the common logic that Freudian psychoanalysis termed "secondary process," which the schizophrenic patient does not employ, adopting instead a logic that is primitive, private, and autistic.

In his book Arieti criticized Cesare Lombroso's idea of an original connection between artistic genius and mental illness (Lombroso 1894), as well as the psychoanalytic theories about creativity expounded by Freud and Jung that focused on unconscious motivation—sexual drives (Freud) or archetypes (Jung). Instead, Arieti suggested that creativity results from a combination of two basic mechanisms of thought—those pertaining to the primary process, that is, the unconscious; and those pertaining to the secondary process, that is, logical thinking. As Kandinsky had pointed out (Kandinsky 1947), creativity had to do with the merging of an individual's "inner space" (the unconscious of the primary process) and his reasoning (the secondary process). To describe this combination, Arieti proposed the term "tertiary process," providing some examples of what he meant. One was the work of Marc Chagall:

Chagall's paintings do not represent the ordinary aspects of things. They disregard the accident of reality. They respect neither anatomy nor perspective, nor the laws of gravity, nor those of space and time. Often animals and monsters, as well as simple human folk, are airborn or put together in unnatural relations. All this is primary process; all this bears resemblance to schizophrenia. The resemblance, however, ends here; and here Chagall's miracle starts. His language, while apparently abstruse, is actually understood the world over ... The Chagallian transformation permits us to recognize the world and see new meanings in it. ... In this disorder we recognize a hidden order—the hidden order of the tertiary process ... It is a new combination of abstract concepts expressed by visual forms and aesthetic relations. The mixture of concepts is the celebration of love and joy. The fish jump out of the sea and human beings rotate in the air in circles of jubilation. We recognize aesthetic devices that give secondary-process structure to primary-process spontaneous organization (Arieti 1976, pp. 223–226).

Another example was Dali's drawing of a hamlet. According to Arieti, this work showed clearly that the recognition of similarities in different and apparently unconnected things was a basic feature of creativity:

The drawing portrays a group of homes, a small village, protected by predatory birds (the rest of the world). A tree in the middle beautifies this little oasis of love and harmony. However, when we look at the drawing in its totality, we discover that it represents not a hamlet but man himself. We may distinguish the vertebral column, trunk, ribs, arms, and legs. Man is identified with the hamlet because of the similar shape; each part of the hamlet is identified with a similarly-shaped part of the human form. ... Here Von Domarus' principle is applied visually. We do not deal with identical predicates but with visually similar parts that lead to the identification. ... This phenomenon of the second or third image occurs frequently in Dali's works. ... Dali correctly believes that paranoiacs have "a special capacity for the recognition of double images inasmuch as their disordered minds are hypersensitive to hidden appearances, real or imagined." We must agree with him. The admirer of Dali's paintings is led by the artist to discover these similarities that he would not be able to notice by

> himself. Is Dali paranoiac or paranoid? Not in a clinical sense. As he expresses himself in his writings, he has "his own paranoia." I interpret his words as meaning that he has a very unusual accessibility to the primary process. But this is the artistic power of Dali: that he retains complete contact with the secondary process, so that the secondary process is able to control the primary. This control is well demonstrated in his paintings, which disclose an overall pattern of exactitude superimposed on an absurd content. This almost photographic exactitude ... contrasts strikingly with the absurdity of the content, but it also mingles with it in an unparalleled way ... I believe that psychiatry, and the psychology of the creative process, owe a debt of gratitude to Salvador Dali (Arieti 1976, pp. 229–231).

Here we understand clearly the relevance of Arieti's study of schizophrenia for his analysis of creativity, since the ability to grasp similarities and unexpected connections in the surrounding world, typical of the creative person, was also a feature of schizophrenic thought. In fact, the recognition of common characteristics in things without any evident connection was a feature of paleological thinking, which was founded on identification—a process that also applied to art. In this regard, the use of metaphors in poetry was enlightening. Borrowing from Elizabeth Drew's interpretation of "The Sick Rose" by William Blake (Drew 1959), Arieti wrote:

> Ostensibly the poem is about a beautiful flower that has been invaded and soon will be destroyed by an ugly worm. But there are many more levels of metaphorical meanings. What comes easily to mind is that the rose stands for a beautiful woman and that the worm stands for a fatal illness that soon will destroy her. In fact, the poet addresses the rose as a person. ... In the poem we see the woman *in* the rose. The woman and the rose are fused; but it is not that bizarre fusion that we see in schizophrenic drawings and delusions. The woman and the rose, though fused, retain their individuality. The retention of their individuality permits a comparison, yet does not lead to identification. ... By putting the sick rose and the sick woman together, we become at least partially *conscious* of a class: the class of "beautiful life destroyed by illness." ... The primary process tends to remain primary: the rose and the woman tend to interchange ..., but as the concept of the class "beautiful life destroyed by illness" emerges, their fusion does not become, so to speak, consummated. They remain distinct. ... Here, thus, is an important difference between art and psychopathology: *Whereas in the psychopathologic use of the primary process there is no consciousness of abstraction, in art the use of the primary process does not eliminate the abstract. On the contrary, it is through the medium of the primary process that the abstract concept emerges.* The poet discovers that things abound in similarities. New similarities take on new meanings because each recognized similarity is a concept and implies the formation of a new class. One of the main ways of expanding knowledge that the aesthetic fields have in common with science is this formation of new classes or categories (Arieti 1976, pp. 137–139).

Here we are well beyond a purely aesthetic pleasure, since words are capable of creating new conceptual horizons to enrich the world, broadening and deepening our knowledge. In this sense, according to Arieti, art has ultimately a "universalist" function, since it does not concern only the individual's experience, but the human experience as a whole.

The capability of poetry to create new concepts and to show us new worlds is also a characteristic of science, as mathematician Henri Poincaré pointed out in a study that became a classic in the literature about creativity (Poincaré 1913). It might seem strange, Arieti remarked, to compare two different realms like poetry—with its suggestive and elusive language that constantly oscillates between primary and secondary processes—and science—with its clear, unambiguous language, where no traces can be found of the primary process. And yet, the process of creativity is similar, as both poetry and science discover connections between things never before related. The concept of universal gravitation, for example, stemmed from recognizing a similarity between two forces—one that causes an apple to fall to the earth, and another that keeps heavenly bodies in their orbits. Another example of the role played by similarities in scientific discoveries was Darwin's theory of natural selection, which recognized a relationship between two different fields of knowledge—the demographic observations of Malthus and biology.

The parallel between poetry and science is undoubtedly one of the more innovative features of Arieti's book, one that did not escape the notice of Ferruccio Di Cori, who, introducing Arieti to the audience at the Sigmund Freud Award ceremony in 1978, defined him as a romantic scientist: "Arieti is essentially a romantic, and underneath his romanticism there is sheer poetry—poetry associated and linked with science. … He is a psychiatrist who loves the quintessence of scientific search and, at the end, he lands in poetry" (SAP 1978c). In his speech Di Cori also described his colleague as an *agent provocateur,* capable of renewing his challenge to science, "day after day, year after year" (SAP 1978c).

Indeed, *Creativity* represents a clear and meaningful example of Arieti's challenge to science. On the one hand, his detailed analysis of the mechanisms involved in the creative process epitomized his search for a scientific "dissection" of this phenomenon (Arieti 1977b, p. 3) along a path opened by the study of schizophrenic mental processes, which he conceived as "laboratory experiments" (Arieti 1976, p. 412); thus, creativity could be described as a process in which the basic mechanisms of human cognition (identification and distinction) dialectically relate, integrate, and sustain one another, revealing new and unexplored similarities and thus enriching our knowledge. On the other hand, Arieti's study of creativity seems to transgress the bounds of scientific inquiry, and the core of his challenge to science can be found in the use of the word "Magic" in the title of his book—for a scientific "dissection" of the creative process cannot wholly explain creativity, which goes beyond matter and phenomenal reality. As Arieti stated clearly, art, helping us to reach a spiritual level in life, can be looked at as "the humble human counterpart of God's creation" (Arieti 1976, p. 4). In art, he wrote in his notes, "the physical and the spiritual coincide" and there is an "union of finitude with infinity" (SAP Ub). It is a "new synthesis" of "transcendence and immanence" (SAP Ub) that an anonymous reader compared to the Italian philosopher Benedetto Croce's concept of Spirit (SAP Uf).

With his book Arieti positioned himself between science and religion—a middle ground described as the domain of philosophy (Money-Kyrle 1958). Arieti did not, however, aim at a self-referential "system of abstract thoughts that the philosophers themselves have created" (Arieti 1976, p. 288). Rather, he conceived philosophy as a "science of similarities," along the lines of von Bertalanffy's General System Theory:

> Von Bertalanffy's tertiary process consisted of finding isomorphies or similarities among the various fields, or systems, and in founding the science of similarities. These isomorphies became the principles or laws of general system theory. But in the words of Boulding, general system theory is itself "the skeleton of science in the sense that it aims to provide a framework

> or structure of systems on which to hang the flesh and blood of particular disciplines and particular subject matters in an orderly and coherent corpus of knowledge (Arieti 1976, pp. 289–290).

Undoubtedly, Arieti's book is itself a creative undertaking, if nothing else for its challenge to science. The writing of *Creativity* can perhaps be considered the outcome of Arieti's own tertiary process, as he suggested in the dedication of the book to Ludwig von Bertalanffy and Luisa Orvieto, his teacher at the elementary school in Pisa—"two nourishments of the mind" placed "one in the intellect, the other close to heart" (Arieti 1976, p. vii) from the synthesis of which his book took shape.

The dedication to von Bertalanffy and Orvieto shows one of the many relational and experiential elements that contributed to the book, in which Arieti, as acutely grasped by Primo Levi, conveyed not only the results of his study but also "a huge life experience" (SAP 1979a, my translation)—from his love for painting, which he shared with Marianne, herself "a dedicated painter" (SAP Ug), to his love for literature and writing (SAPa; SAPh); from his passion for music and especially opera, productions of which he attended regularly (SAP Ug), to his interest in the Classics, which brought him closer to his son James (SAPa; Arieti & Arieti). These interests led him to new encounters and new relationships; for example, with Jacob Landau, an American artist whose work Arieti appreciated for its exploration of the human condition (SAPc); and with Christopher Roberts, a scholar of musical traditions in primitive cultures, with whom he had a rich and dense correspondence (SAPa). Furthermore, other perspectives opened up to Arieti when he was working on his book—Zerka Moreno's psychodrama, for example, which he started to explore in these years (SAP 1976a, 1976b).

Creativity was published in 1976, just one year after Rollo May's *The Courage to Create* (May 1975). The two books were very different, beginning with their styles. The work of May, written in an existentialistic outlook reminiscent of Protestant theologian Paul Tillich's *The Courage To Be* (Tillich 1952), was simpler to read than Arieti's book, which was characterized by the attempt to combine humanities and science. This difference was highlighted in Dannye Romine's review, which described *Creativity* "weighty both in content and size" (Romine 1976, p. 6F) compared to May's book.

Despite good sales and many good reviews (SAPd; SAPe), some scathing reviews did appear. Someone objected to Arieti's encroachment on fields other than his own (SAP 1976g). Another complained that "Arieti is like a tone-deaf listener at a concert" (Lacy 1976, p. 15); another wrote, "Arieti's theory seems not only anachronistic but verbose" (Peterson 1976, p. 87). And so on.

A very bad review appeared in *The New York Times Book Review,* written by literary and theater critic Richard Gilman, who called Arieti's discussion of aesthetics "naive, predictable, derivative and in some instances obtuse" (Gilman 1976, p. 173). In particular, Gilman faulted Arieti for his distortion of the creative process by means of its scientific dissection. "Everything in studies of this kind tends toward abstraction, abstraction of an especially pallid, pompous and empty kind. … Art doesn't provide the sort of evidence science requires" (Gilman 1976, p. 173).

Reading this review during his summer vacation in Sardinia left Arieti quite shaken. His secretary, Joan Kirtland, writing to him from New York, tried to reassure him by reporting a message from his publisher: "Tell him I urge him not to be overly concerned about the review. The review is so strongly worded that the readers will easily see that is not objective" (SAP 1976e). The reassurance had only a mild effect, and Arieti decided to address a response to the editor of *The New York Times Book Review,* Harvey Shapiro:

> A reader refractory to the psychological approach apparently has difficulty in understanding a psychological book. My book is a psychological book, not a treatise on aesthetics. It attempts to explain the psychological processes which go on in the act of creativity ... The act of creativity can be studied with many approaches, including the scientific. Gilman believes that the best words on the subject are those written by philosophers and artists themselves. It may be so, but other approaches should not be discouraged (SAP 1976f)

The psychiatrist David Forrest wrote in defense of Arieti that the core of the critique was the reviewer's unwillingness to accept the "scientific humanism" (SAP 1976d) that characterized the book. In short, Arieti's analysis, "more Apollonian than Dionysian" (SAP 1976d), could be neither endorsed nor welcomed by those who deemed it necessary to put science aside to explore the deepest secrets of creativity.

Harshly criticized by art and literary critics, the book was nevertheless well received by psychologists and psychiatrists. It was enthusiastically reviewed by Paul Torrance in *Contemporary Psychology* (Torrance 1977) as well as by Judd Marmor in *The American Journal of Psychiatry* (Marmor 1976). Nevertheless, Judd Marmor, despite finding Arieti's scientific analysis of creativity fully convincing, pointed out questions raised in the book that remained unanswered, questions encapsulated in the title's reference to magic. The same point appeared in Marjorie Meehan's review in *The Journal of the American Medical Association*, where she noted that, despite Arieti's digging into neurology, neuropathology and general system theory, ultimately art remains a mystery that science has not successfully explained (Meehan 1976).

Undoubtedly, with its reference to the spiritual dimension, *Creativity* urged scientists, so to speak, to look at what is "secret," or at least to acknowledge its existence—an exhortation that medical and psychiatric audiences could hardly embrace. The culture of American physicians "does not foster the kind of learning that your papers and your books demonstrate," wrote a member of the public who attended a conference on art where Arieti had received a tepid response from an audience unappreciative of the scope of his provocative thinking (SAP 1980b).

It is perhaps also for this reason that, looking beyond psychiatry, in 1980, Arieti decided to "become involved in other activities" (SAP 1980a). Chief among these was organizing a Foundation for the Study and Cultivation of Creativity, to which Arieti devoted much of his energy, searching for funding and establishing new international relationships (SAPa; SAPi).[11]

[11] More particularly, Arieti communicated with Carlo Brumat, of the French Institut Européen d'Administration des Affaires, who invited him to participate in various seminars on creativity and innovation in Europe.

In the summer of 1980, the Foundation was eventually incorporated as a non-profit organization. Its initial aim was "to develop communications among creative people, students of creativity and their colleagues in other disciplines in the sciences, the humanities, education, and whatever fields enhance innovations in human society" (SAP 1980c). At the same time, Arieti started to work on a new book on the goals of the Foundation, which unfortunately remained unfinished and was never published (SAPb; Figs. 6.3 and 6.4).

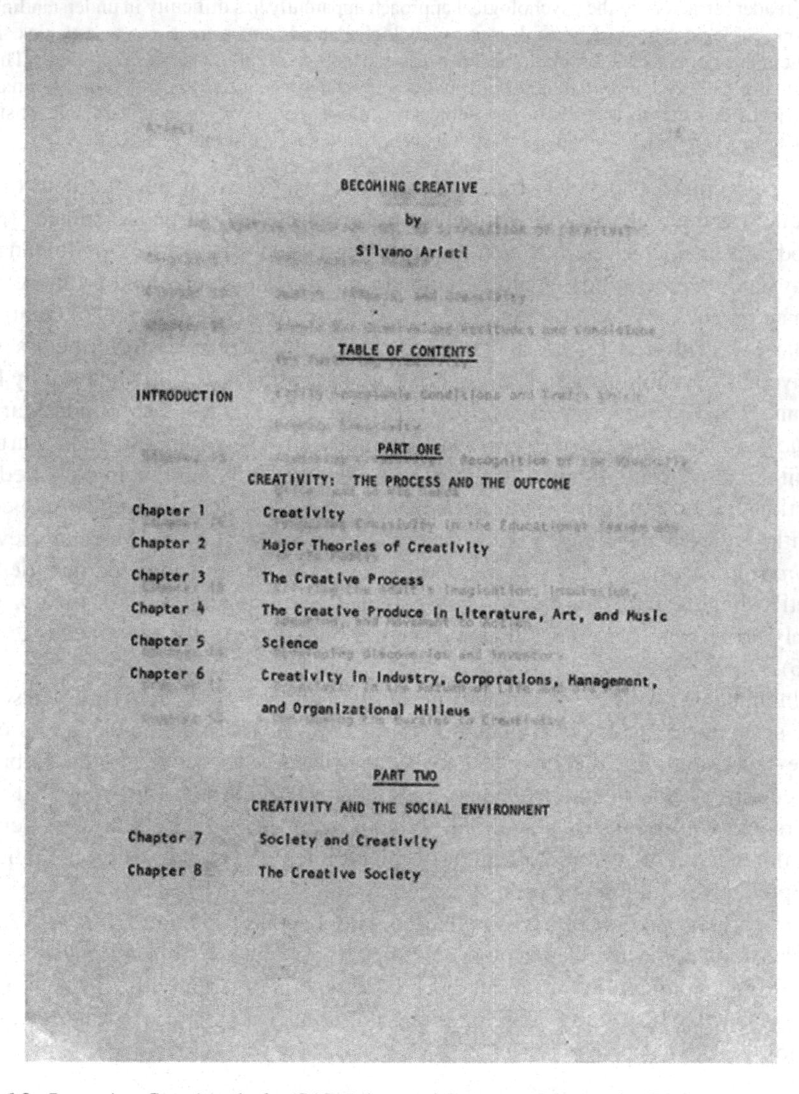
BECOMING CREATIVE

by

Silvano Arieti

TABLE OF CONTENTS

INTRODUCTION

PART ONE

CREATIVITY: THE PROCESS AND THE OUTCOME

Chapter 1 Creativity
Chapter 2 Major Theories of Creativity
Chapter 3 The Creative Process
Chapter 4 The Creative Produce in Literature, Art, and Music
Chapter 5 Science
Chapter 6 Creativity in Industry, Corporations, Management, and Organizational Milieus

PART TWO

CREATIVITY AND THE SOCIAL ENVIRONMENT

Chapter 7 Society and Creativity
Chapter 8 The Creative Society

Fig. 6.3 *Becoming Creative*, drafts (SAP, Library of Congress, Manuscript Division) © Courtesy of James Arieti—All rights reserved

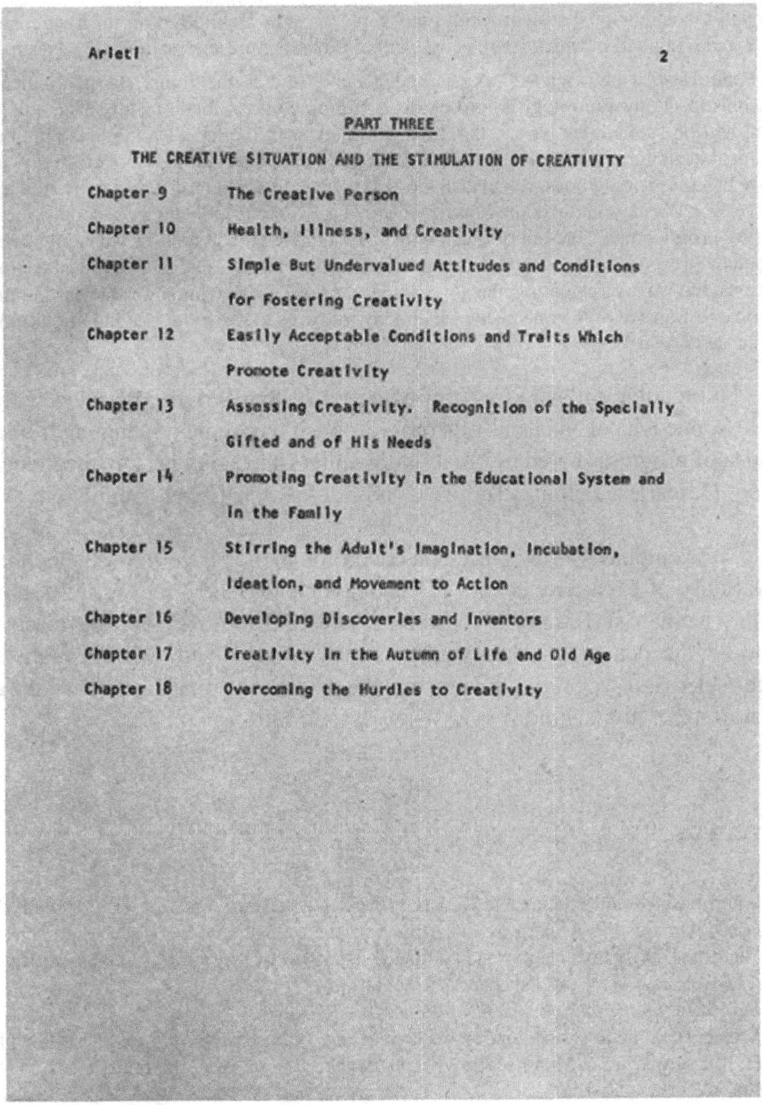

Arieti 2

PART THREE

THE CREATIVE SITUATION AND THE STIMULATION OF CREATIVITY

Fig. 6.4 *Becoming Creative*, drafts (SAP, Library of Congress, Manuscript Division) © Courtesy of James Arieti—All rights reserved

With his work to foster creativity Arieti hoped to counteract the cultural and social crises of the day. In his proposal for the Foundation he wrote:

> Every human being is potentially creative. Unlike other species of animals ..., each human being is potentially an innovator ... There is a great deal to learn about the enhancement of creativity. ... Unfortunately, the cultural climate of our day predisposes [us] to uniformity, conformity, lack of initiative, and discouragement of autonomy. ... The main obstacle comes

through the educational system. Until [age] three or four, a child has an inherent creative drive and imagination which later become inhibited by the present methods of education.

Futurology, a new science which predicts and studies the future, is predominantly pessimistic. Many futurologists foresee overwhelming crises for humanity. The environment will be spoiled, the air and the water polluted; with the increase in population there will not be enough space to live, not enough oxygen to breathe. A renewal of creativity must prove that these predictions, worse than those described by Orwell in *1984* and by Huxley in *Brave New World*, will not come true. ... We are at a crossroads. We either become a society that is "programmed" and out of touch with ... its humanistic dreams, or we recognize our imperative responsibility to nurture imagination and creativity. The media, the educational systems, the ways of parenting, the governmental priorities, the funding vehicles—all must become responsive of the one factor which may allow us to resist the graying of our global existence (SAP 1980d).

The Foundation embodied a form of activism that Arieti connected with both "the Messianic idea of Judaism" of a progressive improvement in human life and the lofty idea of progress shared by great philosophers and scientists of "the modern era (Galileo, Descartes, Leibniz, Turgot, Condorcet, Kant, Hegel, Comte, etc.)" (SAP Uh).

Also, the emphasis Arieti placed on education in the platform for the Foundation for the Study of Creativity echoes the endeavor undertaken by Maria Montessori, herself a psychiatrist (Kramer 1976; Babini & Lama 2000). Thus, while re-affirming, like she did, the idea that psychiatry must venture beyond itself, Arieti brought human science back to its own origins, relaunching its challenge—that of a knowledge aimed at the welfare of humankind.

References

APA Committee on Nomenclature and Statistics (1952). *Diagnostic and Statistical Manual. Mental Disorders.* American Psychiatric Association.

APA Committee on Nomenclature and Statistics (1968). *Diagnostic and Statistical Manual of Mental Disorders.* American Psychiatric Association.

Arieti, S. (1955). *Interpretation of Schizophrenia.* Basic Books.

Arieti, S. (1959). Some Socio-Cultural Aspects of Manic Depressive Psychoses and Schizophrenia. In J.H. Masserman & J. Moreno, *Progress in Psychotherapy. Vol. IV* (pp.140–152). Grune & Stratton.

Arieti, S. (1960). Etiological Considerations of Schizophrenia. In S.C. Scher & H.R. Davis, *The Out-Patient Treatment of Schizophrenia* (pp. 24–29). Grune & Stratton.

Arieti, S. (1964). The Rise of Creativity: From Primary to Tertiary Process. *Contemporary Psychoanalysis,* 1 (1), 51–69.

Arieti, S. (1966). Creativity and Its Cultivation: Relation to Psychopathology and Mental Health. In S. Arieti (Ed.), *American Handbook of Psychiatry. Vol. III* (pp. 722–741). Basic Books.

Arieti, S. (1967). *The Intrapsychic Self: Feeling, Cognition and Creativity in Health and Mental Illness.* Basic Books.

Arieti, S. (1968). The Present Status of Psychiatric Theory. *The American Journal of Psychiatry,* 124 (12), 1630–1639.

Arieti, S. (1969). Presidential Message. *Newsletter. Society of Medical Psychoanalysts,* 11 (3), 29–30.

Arieti, S. (1971). Current Ideas on the Problem of Psychosis. In P. Doucet & C. Laurin, *Problems of Psychosis. International Colloquium on Psychosis. Montreal, 5–8 November 1969* (pp. 3–21). Excerpta Medica.

Arieti, S. (1973a). Critical Evaluations: Concepts of Schizophrenia. An Abnormal Way of Dealing with an Abnormal Situation. *The Canadian Psychiatric Association Journal,* 18 (3), 253–254.

Arieti, S. (1973b). Editorial. *The Journal of the American Academy of Psychoanalysis,* 1 (1), 1–2.

Arieti, S. (1973c). Schizophrenic Art and Its Relationship to Modern Art. *The Journal of the American Academy of Psychoanalysis,* 1 (4), 333–365.

Arieti, S. (1974a). *Interpretation of Schizophrenia.* Second Edition. Basic Books.

Arieti, S. (1974b). The Mother and Father of the Schizophrenic: A Reconsideration. In G. Chrzanowski, A. Heigl-Evers, H. V. Brazil, & W. Schwidder, *Recent Developments in Psychoanalysis. Theory and Practice. Vol. VI. The Individual, the Family and Society in the Tension Between Freedom and Coercion* (pp. 439–448). International Forum of Psychoanalysis.

Arieti, S. (1974c). An Overview of Schizophrenia from a Predominantly Psychological Approach. *The American Journal of Psychiatry,* 131 (3), 241–248.

Arieti, S. (1975a). Dr. Silvano Arieti's Response. *The Journal of the American Academy of Psychoanalysis,* 3 (3), 234.

Arieti, S. (1975b). Psychiatric Controversy: Man's Ethical Dimension. *The American Journal of Psychiatry,* 132 (1), 39–42.

Arieti, S. (1975c). *Psichiatria e oltre.* Il Pensiero Scientifico.

Arieti, S. (1976). *Creativity: The Magic Synthesis.* Basic Books

Arieti, S. (1977a). Parents of the Schizophrenic Patient: A Reconsideration. *The Journal of the American Academy of Psychoanalysis,* 5 (3), 347–358.

Arieti, S. (1977b). *New Views of Creativity.* Geigy Pharmaceuticals.

Arieti, S. (1980a). Depressione grave. *Rivista di psichiatria,* 15 (1), 21–25.

Arieti, S. (1980b). New Psychological Approaches to Creativity. *Contemporary Psychoanalysis,* 16 (3), 287–306.

Arieti, S., Arieti, J. (1977). *Love Can Be Found.* Harcourt Brace Jovanovich.

Arieti, S., Bemporad, J. (1978). *Severe and Mild Depression. The Psychotherapeutic Approach.* Basic Books.

Arieti, S., Lorraine, S. (1972). The Therapeutic Assistant in Treating the Psychotic. *International Journal of Psychiatry,* 10 (3), 7–22.

Babini, V.P., Lama, L. (2000). *Una "donna nuova". Il femminismo scientifico di Maria Montessori.* Franco Angeli.

Balbuena Rivera, F. (2016). The Relevance of Arieti's Work in the Age of Medicalization. *The American Journal of Psychoanalysis,* 76 (3), 266–280.

Bateson, G., Jackson, D. D., Haley, J., & Weakland, J. (1956). Toward A Theory of Schizophrenia. *Behavioral Science,* 1 (4), 251–264.

Bernard, C. (1865). *Introduction à l'étude de la médecine expérimentale.* Baillière.

Campbell, N. D. (2014). The Spirit of St. Louis: The Contributions of Lee N. Robins to North American Psychiatric Epidemiology. *International Journal of Epidemiology,* 43 (Suppl. 1), 19–28.

Clayton, P. J. (2006). Training at Washington University School of Medicine in Psychiatry in Late 1950's, From the Perspective of an Affective Disorders Researcher. *Journal of Affective Disorders,* 92 (1), 13–17.

Decker, H. (2007). How Kraepelinian Was Kraepelin? How Kraepelinian Are the Neo-Kraepelinians? From Emil Krapelin to DSM-III. *History of Psychiatry,* 18 (71), 337–360.

Decker, H. S. (2013). *The Making of DSM-III: A Diagnostic Manual's Conquest of American Psychiatry.* Oxford University Press.

Disertori, B. (1947). *Il libro della vita.* Mondadori.

Drew, E. (1959). *Poetry: A Modern Guide to Its Understanding and Enjoyment.* Dell Publishing Co.

Eckardt, M. (1977). The Nature of Our Knowledge. *The Journal of the American Academy of Psychoanalysis,* 5 (4), 415–417.
Edel, L. (1975). The Future of Humanism. *The Journal of the American Academy of Psychoanalysis,* 3 (1), 5–20.
Ehrenberg, A. (1998). *La fatigue d'être soi. Dépression et société.* Editions Odile Jacob.
Feighner, J., Robins, E., Guze, S., Woodruff, R., Winokur, G., & Munoz, R. (1972). Diagnostic Criteria for Use in Psychiatric Research. *Archives of General Psychiatry,* 26 (1), 57–63.
Fischer, B. (2012). A Review of American Psychiatry Through Its Diagnoses. The History and Development of the Diagnostic and Statistical Manual of Mental Disorders. *The Journal of Nervous and Mental Disease,* 200 (12), 1022–1030.
Gilman, R. (1976, August 1). An Analyst Analyzes Artists and Art. Creativity: The Magic Synthesis by Silvano Arieti. *The New York Times Book Review,* 173.
Guntrip, H. (1973). Science, Psychodynamic Reality and Autistic Thinking. *The Journal of the American Academy of Psychoanalysis,* 1 (1), 3–22.
Healy, D. (1997). *The Anti-Depressant Era.* Harvard University Press.
Healy, D. (2002). *The Creation of Psychopharmacology.* Harvard University Press.
Herzberg, D. L. (2010). *Happy Pills in America. From Miltown to Prozac.* Johns Hopkins University Press.
Kalem, T. E. (1975, April 28). Cash and Culture. *Time,* 105, 95.
Kandinsky, V. V. (1947). *Concerning the Spiritual in Art.* Wittenborn.
Kendler, K. S., Munoz, R., & Murphy, G. (2010). The Development of Feighner Criteria: A Historical Perspective. *The American Journal of Psychiatry,* 167 (2), 134–142.
Kirk, S. A., Kutchins, H. (1992). *The Selling of DSM: The Rethoric of Science in Psychiatry.* Routledge.
Kline, N. (1974). *From Sad to Glad: Kline On Depression.* G.P. Putnam's Sons.
Kramer, R. (1976). *Maria Montessori: A Biography.* Putnam.
Kuhn, T. (1962). *The Structure of Scientific Revolutions.* University of Chicago Press.
Lacy, A. (1976.) To Bring the New Into Being. Creativity: The Magic Synthesis by Silvano Arieti. *The Chronicle of Higher Education,* 13, 15.
Laing, R. (1960). *The Divided Self.* Tavistok.
Laing, R. (1967). *The Politics of Experience.* Pantheon Books.
Lamb, S. D. (2014). *Pathologist of the Mind: Adolf Meyer and the Origins of American Psychiatry.* Johns Hopkins University Press.
Lidz, T. (1966). Adolf Meyer and the Development of American Psychiatry. *The American Journal of Psychiatry,* 123 (3), 320–332.
Lombroso, C. (1894). *L'uomo di genio in rapporto alla psichiatria, alla storia ed all'estetica.* Fratelli Bocca.
Marmor, J. (1976). Creativity: The Magic Synthesis by Silvano Arieti. *The American Journal of Psychiatry,* 133 (12), 1473.
May, R. (1975). *The Courage to Create.* Norton.
McReynolds, P. (1975). The World of Schizophrenia. *Contemporary Psychology,* 20 (7), 548–550.
Meehan, M. (1976). Creativity: The Magic Synthesis by Silvano Arieti. *The Journal of the American Medical Association,* 236 (16), 1893–1894.
Menninger, K., Mayman, M., & Pruyser, P. (1963). *The Vital Balance: The Life Process in Mental Health and Illness.* Viking Press.
Money-Kyrle, R. (1958). Psycho-Analysis and Philosophy. In J.D. Sutherland (Ed.), *Psychoanalysis and Contemporary Thought* (pp. 102–123). Hogarth Press.
Peterson, D. (1976). Creativity: The Magic Synthesis by Silvano Arieti. *The Atlantic Monthly* 238 (August), 86–87.
Poincaré, H. (1913). *The Foundations of Science.* The Science Press.
Redlich, F. (1974). Psychoanalysis and the Medical Model. *The Journal of the American Academy of Psychoanalysis,* 2 (2) 147–158.
Riesman, D., Glazer, N., & Denney, R. (1950). *The Lonely Crowd.* Yale University Press.

Rifkin, A. H. (1974). Science, Anti-Science and the Apotheosis of Unreason. *The Journal of the American Academy of Psychoanalysis*, 2 (3), 181–186.

Romine, D. (1976, June 27). Two New Books on Creativity. *Charlotte Observer*, 6F.

Sabshin, M. (1990). Turning Points in Twentieth-Century American Psychiatry. *The American Journal of Psychiatry*, 147 (10), 1267–1274.

SAPa. *Correspondence, 1940–1981.*

SAPb. *Speeches and Writings, 1940–1981. Becoming Creative, 1980.*

SAPc. *Speeches and Writings, 1940–1981. Collected material. Jacob Landau, 1970–1976.*

SAPd. *Speeches and Writings, 1940–1981. Creativity: The Magic Synthesis. Correspondence, 1975–1978.*

SAPe. *Speeches and Writings, 1940–1981. Creativity: The Magic Synthesis. Reviews, 1976–1977.*

SAPf. *Speeches and Writings, 1940–1981. Interpretation of Schizophrenia. National Book Award, 1975.*

SAPg. *Speeches and Writings, 1940–1981. Interpretation of Schizophrenia. Second Edition. Reviews, 1974–1975.*

SAPh. *Speeches and Writings, 1940–1981. Plays.*

SAPi. *Subject File, 1914–1981. Organizations. Foundation for the Study and Cultivation of Creativity, 1980–1981.*

SAPj. *Subject File, 1914–1981. Teaching. Psychotherapy of Schizophrenia Course, 1975–1978.*

SAP (1956a). L, December 30 (Silvano Arieti to David Reisman). In *Correspondence, 1940–1981.*

SAP (1956b). T (Constitution and By Laws of the American Academy of Psychoanalysis). In *Subject File, 1914–1981. Organizations. American Academy of Psychoanalysis, 1970–1980.*

SAP (1957a). L, January 2 (David Reisman to Silvano Arieti). In *Correspondence, 1940–1981.*

SAP (1957b). L, May 27 (Silvano Arieti to David Riesman). In *Correspondence, 1940–1981.*

SAP (1957c). L, June 8 (David Riesman to Silvano Arieti). In *Correspondence, 1940–1981.*

SAP (1957d). T (Silvano Arieti, *The Decline of Manic-Depressive Psychoses: Its Significance in the Light of Dynamic and Social Psychiatry*). In *Speeches and Writings, 1940–1981. Lectures.*

SAP (1961a). L, January 30 (Silvano Arieti to Rudolph Arnheim). In *Correspondence, 1940–1981.*

SAP (1961b). L, February 4 (Rudolph Arnheim to Silvano Arieti). In *Correspondence, 1940–1981.*

SAP (1961c). L, February 18 (Silvano Arieti to Rudolph Arnheim). In *Correspondence, 1940–1981.*

SAP (1962). L, September 22 (Silvano Arieti to Alexander Mathé). In *Correspondence, 1940–1981.*

SAP (1964). T, May 27 (Silvano Arieti, *The Rise of Creativity: From Psychopathology to Innovation*). In *Speeches and Writings, 1940–1981. Lectures.*

SAP (1967). L, May 31 (Beppino Disertori to Silvano Arieti). In *Speeches and Writings, 1940–1981. Articles.*

SAP (1968a). L, June 18 (Silvano Arieti to Henry Brill).

SAP (1968b). L, August 19 (Silvano Arieti to Henry Brill)

SAP (1969a). L, November 30 (Silvano Arieti to Loren Mosher).

SAP (1969b). L, December 12 (Silvano Arieti to John Broogher).

SAP (1969c). L, December 23 (Clark J. Bailey to Silvano Arieti).

SAP (1970). T, January 28 (Silvano Arieti, *The Concept of Schizophrenia*). In *Speeches and Writings, 1940–1981. Lectures.*

SAP (1971a). L, March 1 (John Schimel to Gardner Spurgin). In *Subject File, 1914–1981. Organizations. American Academy of Psychoanalysis, 1970–1980.*

SAP (1971b). L, April 16 (John Schimel to M. Harris). In *Subject File, 1914–1981. Organizations. American Academy of Psychoanalysis, 1970–1980.*

SAP (1971c). L, June 25 (Irving Bieber to Silvano Arieti). In *Subject File, 1914–1981. Organizations. American Academy of Psychoanalysis, 1970–1980.*

SAP (1971d). L, November 15 (Silvano Arieti to Harry Guntrip). In *Subject File, 1914–1981. Organizations. American Academy of Psychoanalysis, 1970–1980.*

SAP (1971e). L, November 24 (Harry Guntrip to Silvano Arieti). In *Subject File, 1914–1981. Organizations. American Academy of Psychoanalysis, 1970–1980.*

SAP (1972a). L, January 2 (Silvano Arieti to Theodore Lidz). In *Subject File, 1914–1981. Organizations. American Academy of Psychoanalysis, 1970–1980.*
SAP (1972b). L, March 6 (Walter Bonime to Silvano Arieti). In *Correspondence, 1940–1981.*
SAP (1972c). L, March 15 (Silvano Arieti to Gardner Spurgin). In *Subject File, 1914–1981. Organizations. American Academy of Psychoanalysis, 1970–1980.*
SAP (1972d). L, April 10 (Harry Guntrip to Silvano Arieti). In *Subject File, 1914–1981. Organizations. American Academy of Psychoanalysis, 1970–1980.*
SAP (1972e). L, June 12 (Francis Braceland to Silvano Arieti). In *Subject File, 1914–1981. World Biennial of Psychiatry and Psychotherapy, editor. Correspondence.*
SAP (1973a). L, April 24 (Harry Guntrip to Silvano Arieti). In *Subject File, 1914–1981. Organizations. American Academy of Psychoanalysis, 1970–1980.*
SAP (1973b). L, September 18 (Silvano Arieti to the Members of the American Academy of Psychoanalysis). In *Subject File, 1914–1981. Organizations. American Academy of Psychoanalysis, 1970–1980.*
SAP (1973c). L, September 25 (Silvano Arieti to Iris Topel). In *Correspondence, 1940–1981.*
SAP (1973d). L, November 2 (Silvano Arieti to Iris Topel). In *Correspondence, 1940–1981.*
SAP (1974). L, March 19 (Silvano Arieti to Nathan Segel). In *Correspondence, 1940–1981.*
SAP (1975a). Tmin, April 16 (*National Book Awards 1975a. Advance-Hold for Release*). In *Speeches and Writings, 1940–1981. Interpretation of Schizophrenia. National Book Award, 1975a.*
SAP (1975b). L, April 29 (Silvano Arieti to *New York Medicine*). In *Speeches and Writings, 1940–1981. Interpretation of Schizophrenia. National Book Award, 1975b.*
SAP (1975c). L, May 2 (Francis Braceland to Silvano Arieti). In *Speeches and Writings, 1940–1981. Interpretation of Schizophrenia. National Book Award, 1975c.*
SAP (1975d). T (World Health Organization, *International Conference for the Ninth Revision of the International Classification of Diseases. Chapter V. Mental Disorders*). In *Subject File, 1914–1981. Diagnostic and Statistic Manual of Mental Disorders, 1976.*
SAP (1976a). L, April 1 (Zerka T. Moreno to Silvano Arieti).
SAP (1976b). L, May 18 (Silvano Arieti to Zerka T. Moreno).
SAP (1976c). L, June 30 (Howard Berk to Silvano Arieti). In *Subject File, 1914–1981. Diagnostic and Statistic Manual of Mental Disorders, 1976c.*
SAP (1976d). L, August 11 (David Forrest to *The New York Times Book Review*). In *Speeches and Writings, 1940–1981. Creativity: The Magic Synthesis. Correspondence, 1975–1978.*
SAP (1976e). L, August 19 (Joan Kirtland to Silvano Arieti). In *Speeches and Writings, 1940–1981. Creativity: The Magic Synthesis. Correspondence, 1975–1978.*
SAP (1976f). L, August 19 (Silvano Arieti to Harvey Shapiro). In *Speeches and Writings, 1940–1981. Creativity: The Magic Synthesis. Correspondence, 1975–1978.*
SAP (1976g). L, September 21 (Silvano Arieti to Joshua Abend). In *Speeches and Writings, 1940–1981. Creativity: The Magic Synthesis. Correspondence, 1975–1978.*
SAP (1976h). L, September 21 (Silvano Arieti to Henry Brill).
SAP (1976i). L, October 19 (Silvano Arieti to Simon Nagler). In *Correspondence, 1940–1981.*
SAP (1976j). T *(List of Participants. Conference on Improvements in Psychiatric Classification and Terminology: A Working Conference to Critically Examine DSM-III in Midstream).* In *Subject File, 1914–1981. Diagnostic and Statistic Manual of Mental Disorders, 1976j.*
SAP (1976k). T, Unt. (Remarks by Howard Berk and Hector Jaso at the plenary session of the Conference on DSM in midstream). In *Subject File, 1914–1981. Diagnostic and Statistic Manual of Mental Disorders, 1976k.*
SAP (1976l). T (Silvano Arieti, *Comments by Silvano Arieti About the Report on the Preparation of DSM III*). In *Subject File, 1914–1981. Diagnostic and Statistic Manual of Mental Disorders, 1976k.*
SAP (1977a). L, February 18 (Philip May to Silvano Arieti). In *Correspondence, 1940–1981.*
SAP (1977b). L, March 30 (Leo Jacobs to Silvano Arieti). In *Subject File, 1914–1981. Conferences and Lectures, 1957–1981.*

SAP (1977c). T, April 21 (Silvano Arieti, *New Views of Creativity by Professor Silvano Arieti*). In *Speeches and Writings, 1940–1981. Lectures.*

SAP (1977d). L, December 20 (Silvano Arieti to Gilead Nachmani). In *Correspondence, 1940–1981.*

SAP (1977e). B (*The Forest Hospital Foundation. Medical Models in Psychiatry: Toward a New Synthesis. The 1977d-78 Scientific Lectures Series*). In *Subject File, 1914–1981. Conferences and Lectures, 1957–1981.*

SAP (1978a). L, April 18 (Victor J. Teichner to Silvano Arieti). In *Subject File, 1914–1981. Organizations. American Academy of Psychoanalysis, 1970–1980.*

SAP (1978b). L, June 6 (Silvano Arieti to Simon Nagler). In *Correspondence, 1940–1981.*

SAP (1978c). T, September 22 (Ferruccio Di Cori, *Introductory Speech for the Presentation of the Sigmund Freud Award to Silvano Arieti By the American Association of Psychoanalytic Physicians*). In *Speeches and Writings, 1940–1981. Lectures.*

SAP (1979a). L, February 21 (Primo Levi to Silvano Arieti). In *Correspondence, 1940–1981. Levi, Primo, 1979.*

SAP (1979b). L, February 23 (Silvano Arieti to Clay Dahlberg). In *Subject File, 1914–1981. Organizations. American Academy of Psychoanalysis, 1970–1980.*

SAP (1979c). L, May 18 (Silvano Arieti to Alan Stone). In *Subject File, 1914–1981. Organizations. American Academy of Psychoanalysis, 1970–1980.*

SAP (1979d). L, May 18 (Silvano Arieti to John Spiegel). In *Subject File, 1914–1981. Organizations. American Academy of Psychoanalysis, 1970–1980.*

SAP (1979e). L, May 18 (Silvano Arieti to Jules Masserman). *Subject File, 1914–1981. Organizations. American Academy of Psychoanalysis, 1970–1980.*

SAP (1979f). L, May 18 (Silvano Arieti to Judd Marmor). In *Subject File, 1914–1981. Organizations. American Academy of Psychoanalysis, 1970–1980.*

SAP (1979g). L, May 18 (Silvano Arieti to Robert Gibson). In *Subject File, 1914–1981. Organizations. American Academy of Psychoanalysis, 1970–1980.*

SAP (1979h). L, May 18 (Silvano Arieti to Alfred Freedman). In *Subject File, 1914–1981. Organizations. American Academy of Psychoanalysis, 1970–1980.*

SAP (1979i). L, May 21 (Alan Stone to Silvano Arieti). In *Subject File, 1914–1981. Organizations. American Academy of Psychoanalysis, 1970–1980.*

SAP (1979j). L, May 22 (Robert Gibson to Silvano Arieti). In *Subject File, 1914–1981. Organizations. American Academy of Psychoanalysis, 1970–1980.*

SAP (1979k). L, May 31 (John Spiegel to Silvano Arieti). In *Subject File, 1914–1981. Organizations. American Academy of Psychoanalysis, 1970–1980.*

SAP (1979l). L, June 7 (Jules Masserman to Silvano Arieti). In *Subject File, 1914–1981. Organizations. American Academy of Psychoanalysis, 1970–1980.*

SAP (1979m). L, June 21 (Silvano Arieti to John Nemiah). In *Correspondence, 1940–1981.*

SAP (1979n). L, September 17 (Silvano Arieti to the Fellows of the American Academy of Psychoanalysis). In *Subject File, 1914–1981. Organizations. American Academy of Psychoanalysis, 1970–1980.*

SAP (1980a). L, March 24 (Silvano Arieti to Alfred Yassky). In *Subject File, 1914–1981. Conferences and Lectures, 1957–1981.*

SAP (1980b). L, October 29 (Andrew Rolle to Silvano Arieti). In *Correspondence, 1940–1981.*

SAP (1980c). T (Silvano Arieti, *The Constitution of the Association for the Study and Cultivation of Creativity*). In *Subject File, 1914–1981. Organizations. Foundation for the Study and Cultivation of Creativity, 1980c–1981.*

SAP (1980d). T (Silvano Arieti, *A Proposal for the Foundation for the Study and Cultivation of Creativity*). In *Subject File, 1914–1981. Organizations. Foundation for the Study and Cultivation of Creativity, 1980d–1981.*

SAP (Ua). HN, Unt. (Notes on creativity). In *Subject File, 1914–1981. Conferences and Lectures, 1957–1981.*

SAP (Ub). HN, Unt. (Notes on creativity). In *Speeches and Writings, 1940–1980. The Intrapsychic Self. Notes, circa 1966.*

SAP (Uc). L (Bernard Kaplan to Silvano Arieti). In *Subject File, 1914–1981. American Handbook of Psychiatry, editor. Correspondence, 1956–1980.*

SAP (Ud). PB (Howard P. Rome, *The Main Selection. Silvano Arieti, Jules Bemporad. Severe and Mild Depression. A Review*). In *Speeches and Writings, 1940–1981. Severe and Mild Depression: the Psychotherapeutic approach, circa 1978–1979.*

SAP (Ue). T, Unt. (Notes for a speech on depression). In *Speeches and Writings, 1940–1981. Lectures. Undated.*

SAP (Uf). T (UA, *Alcune idee suscitatemi dalla lettura del libro del Professor Silvano Arieti Creatività, magica sintesi*). In *Speeches and Writings, 1940–1981. Creativity: The Magic Synthesis. Reviews, 1976–1977.*

SAP (Ug). T (Gerard Chrzanowski, *Profile: Silvano Arieti*). In *Subject File, 1914–1981. Personal. Profile, undated.*

SAP (Uh). T (Silvano Arieti, *The Conception of Man*). In *Speeches and Writings, 1940–1980. Articles. Undated.*

Sartorious, N. (Ed.). (1990). *Sources and Traditions of Classification in Psychiatry.* Hogrefe & Huber Pub.

Searles, H. (1958). Positive Feelings in the Relationship Between the Schizophrenic and His Mother. *International Journal of Psychoanalysis,* 39 (6), 569–586.

Shorter, E. (1997). *A History of Psychiatry. From the Era of Asylums to the Age of Prozac.* Wiley.

Shorter, E. (2015). The history of nosology and the rise of the Diagnostic and Statistical Manual of Mental Disorders. *Dialogues in clinical neuroscience,* 17 (1), 59–67.

Slater, E. (1972). Is Psychiatry a Science? Does It Want To Be? *World Medicine,* 8 (1), 79–8.1

Szasz, T. (1957). The Problem of Psychiatric Nosology: A Contribution to a Situational Analysis Of Psychiatric Operations. *The American Journal of Psychiatry,* 114 (5), 405–413.

Szasz, T. (1961). *The Myth of Mental Illness.* Harper.

Tillich, P. (1952). *The Courage to Be.* Yale University Press.

Torrance, P. (1977). Synthetizing About The Magic Synthesis. *Contemporary Psychology,* 22 (4), 257–258.

Wilson, M. (1993). DSM-III and the Transformation of American Psychiatry: A History. *The American Journal of Psychiatry,* 150 (3), 399–410.

Wolberg, L. R. (1979). DSM-III and the Taxonomic Stew. *The Journal of the American Academy of Psychoanalysis,* 7 (2), 143–146.

World Wide Medical News Service. (1969). Highlights of Menninger Forum on the Schizophrenic Syndrome. *Roche Report,* 6 (16), 1–3.

Chapter 7
At the End of the Journey

In taking account of the spiritual dimension of human existence, *Creativity* represented an important step in Arieti's search for a "plural rationality" (Moravia 1986) capable of tackling problems generally considered beyond the reach of scientific reasoning.

After the publication of this book, Arieti discussed spirituality in other writings, which, as he pointed out in a letter to Francis Braceland, he intended not only as a philosophical speculation, but as a contribution to psychiatric theory and practice (SAP 1976). As he had done with *Creativity*, he continued to provoke science by addressing issues usually considered outside its jurisdiction.

In "Man's Spirituality and Potential for Creativity as Revealed in Mental Illness," for example, he brought spirituality to the field of medicine. Referring to Claude Bernard, he wrote:

> Extending to psychiatry a concept that Claude Bernard had applied to general medicine, Sigmund Freud interpreted the symptom not just as an expression of illness ..., but also as a phenomenon having "restitutional" dimensions. In other words, at the same time that Freudacknowledged in the symptom a regressive aspect, he saw in it an attempt to regenerate, to heal, to compensate ... The aim of this paper is to show that we can expand further our understanding of the purposefulness or possible use of the symptom, or of the psychological syndrome, ... and recognize in them dimensions that indicate or suggest spirituality and potential for creativity. To demonstrate the creative aspect of the psychological illness is not too difficult ... Much more difficult is demonstrating a spiritual dimension. Some of the difficulties stem from the use of the words "spirituality" and "spiritualism," terms commonly adopted in theological and philosophical circles, but very rarely in the medical or biological literature. ... In the context of this paper, the word "spirituality" does not retain all the possible connections given to it in philosophy, nor does it have mystical implications. It refers to the fact that the psychological symptomatology of some psychiatric conditions can be interpreted not exclusively as a defense or a protection of the self of the patient, but as a process that is directed toward greater common goals, like preserving the dignity of man (Arieti 1980, p. 436).

If the creative process could be illuminated through analysis of the mechanisms of schizophrenic thought, spirituality could be explored through a study of phobias,

R. Passione, *Psychiatry and the Human Condition*, Springer Biographies,
https://doi.org/10.1007/978-3-031-09304-3_7

a subject to which Arieti had given attention at the beginning of his education in psychiatry (Arieti 1961, 1979a).

Unlike schizophrenia, a condition characterized by the patient's withdrawal into a private and autistic world, a phobia is characterized, according to Arieti, by a close connection of the patient to his environment. He does not withdraw from the world but participates in interpersonal relations and social life. He is able to maintain a certain freedom of action, but his existence is severely hindered by a phobia that shapes his whole daily life. Also, according to Arieti, a typical expression of the spiritual dimension of the phobic condition is a "dis-humanization of the source of fear," which he interpreted along the lines of Martin Buber's *I-Thou* philosophy:

> The phobogenic objects may be bridges, cars, high buildings, etc. Animals ... are very often phobogenic objects, but humans are not. If some humans, like policemen or nuns, are experienced as phobogenic, it is by virtue of a special role they play or of the uniforms they wear. Using Martin Buber's terminology, I pointed out that the phobic person who used to experience difficulty in interpersonal relations makes an attempt through his phobias to change the anxiety-provoking *I-Thou* relation into an *I-It*. ...
>
> ... However, when the phobic displaces the source of danger from a person to a thing ..., not only does he do that to diminish his anxiety, but also to protect his fellow human beings whom otherwise he would see in a way unfitting the image of the human being that he has built, retained in his psyche, and which he wants to preserve. It would be horrendous for a human being to be so threatening as the phobogenic can be; only an *It* could be that way. The phobic is the opposite of the slave-owner who reduces a human being to a tool. By resorting to his neurotic mechanism he is able to maintain a dialogic relationship with other persons, even those who caused anxiety. The symbol of evil, hostility, danger, or hate is now an *It*. ... The patient does not become detached, suspicious, or hostile. He becomes afraid of some specific objects or situations but continues to accept the rest of the world. Not only can he maintain a dialogic relation with the threatening person but he may continue to love him or her. ... He is very much involved with people and the world at large. He retains a great capacity and need for love and affection and participate intensely in life, in spite of his fears (Arieti 1980, pp. 437–438).

Arieti's interest in phobias had deep biographical roots. They developed from his acquaintance with Pardo Roques, president (*Parnas*) of the Jewish community in Pisa, whose salon, "with its very special blend of Hebrew intellectuality and passionate internationalism," had been for the young Arieti a place where "the horizons of provincial life, restricted by Fascist nationalism, were made to open out on the great world" (Arieti 1979b, p. 12). Roques suffered from a phobia, and his condition discovered to Arieti the mystery of mental illness, inspiring him to become a psychiatrist. Roques' fear of animals (especially, of dogs) prevented him from escaping Pisa during the German occupation of the town; instead of seeking escape, he closed himself in his house, also providing shelter to eleven fellow citizens who had chosen to stay with him. On 1 August 1944, a squad of German soldiers broke into Pardo's house and killed everyone they found (Arieti 1979b; Forti 1998).

In 1979 Silvano Arieti published *The Parnas,* dedicated to Pardo Roques—an account of the massacre in Pisa. Written from his heart (SAPa), the narrative wove together autobiography, history, and psychiatry, permitting Arieti to investigate, through the story of the victims' last days, the connection of macro- and micro-history that always interested him. In the book, Arieti portrayed the ennobling spiritual content of Roques' mental illness in contrast to the perverse social illness manifested in Nazi barbarity. As Arieti said in a talk at Congregation Shearith Israel in New York City, he tried in this book "to show that in mental illness at times is hidden the spirituality and nobility of the human being and his desire to love and to fight evil through his love of mankind. This is shown by the life of the Parnas" (Gardner 1979, p. 22).

Before its publication, *The Parnas* received endorsements from Elie Wiesel and Primo Levi (SAPc). Wiesel's endorsement appeared on the jacket's cover, Levi's on the inside fold (SAP 1979a, 1979b, 1979e). The book had great success and was listed by *The New York Time Book Review* among the best 1979 titles (SAPc; SAPd; Seebohm 1979). A Pulitzer Prize nominee, it was also considered for a cinematic version (SAP 1979c, 1979d).

As one enthusiastic reviewer noted (Gerchick 1979), a strong religious stamp permeated Arieti's scientific thought in *The Parnas*—highlighting an idea of communion where God and *Thou* blend, as suggested in the book's epigraph from Psalm 23: "I will fear no evil, for *Thou* art with me" (Arieti 1979b, p. vii).

Religiousness as a state of communion with "the other" is a theme underlying the whole corpus of Silvano Arieti. In 1978 Ferruccio Di Cori highlighted this feature of Arieti's thinking at the conclusion of *Interpretation of Schizophrenia*:

> He ends his opus with a quasi-religious statement: "When the person who had the habit of staring vacantly ... focuses on the many little things of life and recognizes sparks here and there and sees again the sun and the stars and the new leaves and hears the rustling of the branches and the children's laughter and craves for what tomorrow will bring, then we believe in greater realizations, then we envision with faith the universality of the human embrace." And that is how he ended, believing in God, his *Interpretation of Schizophrenia* (SAP 1978).

This idea of a "human embrace" emerged also in other writings about psychotherapy, where Arieti calls for an intimate communion of a therapist and his patient (complementing his view of the therapist as an "ambassador of reality") (SAP 1971; Arieti 1971). It appears, too, about social relations, where in *The Will To Be Human* he writes:

> While we are here in the dark and grope and feel our way, let us give each other a sustaining hand. In the short and tortuous path in which we make our unsteady and almost blind steps, let us not push some higher up and others down. Just as individuality is one of the pinnacles of our will, the other is the state of communion which comes after we have striven for the common good. Then Mr. Nobody will be Mr. Somebody, Mr. Everybody, Mr. Man.
>
> Communion, together with love, may compensate for that inherent aloneness that we find in ourselves. Communion includes basic trust ..., which to a large extent we lose early in life. Let us try to reacquire it, not with the naive inexperience of the infant, but with the depth of a mature adulthood. Then your fellow man will not be a person who has the power to use or exploit you, but a peer in a limited and unpredictable reality. Help him and he will help you. Go toward him and he will come toward you. Make him feel happy he has met you. You will be happy you have met him. His hand will not hold a knife; it is open to meet your palm (Arieti 1972, p. 252).

Arieti's religiousness seemed not to be circumscribed by the credo of a specific religion, as noticed by an unidentified correspondent in a 1965 letter: "The night I heard you speak, I saw you as a kind of perfect priest … Suddenly, I wonder, had you not been Jewish, wouldn't you have become a priest?" (SAP 1965).

This religious vein characterizes also Arieti's last book, published in 1981, *Abraham and the Contemporary Mind*—a work that was "not addressed to Jews only," and that was written "by someone who freely confess that he does not practice his religion in an orthodox way" (Arieti 1981a, p. 5).

Dedicated to his two-year-old granddaughter Aviva, the book began with a passage that recalled the lyricism of Arieti's early writings about the sense of infinity and faith aroused by his communion with nature (SAPe; Chap. 2):

> At dusk when the earth seems to quiver, the swallows joyfully circle the air, and in the velvet sky there is an expectation of the first stars. The trees, the bushes, and the grass, thirsty for dew, welcome the arriving night with their fragrances. Once at such an hour, three to four thousand years ago, between spring and summer, in an obscure part of the Near Orient, a forty-eight-year-old man ... feels as if he is being penetrated by the glory of all that he can see, hear, smell, and touch, and by the rest of the universe that he can think of and imagine. Like his contemporaries ..., this man ... feels possessed by that which surrounds him ... But then suddenly and yet calmly, in recognition of something that has been growing inside him for a long time, he feels he must for a few moments at least silence all earthly sounds, close his eyes to the glory of all celestial and terrestrial colors ... And as he does so, his ephemeral grasp touches the eternal: with his inner ear he can hear the eternal voice, the invisible appears to him in an invisible way, and the unthinkable becomes part of his thoughts. Even all the echoes of the human songs of sorrow and joy fade away and are replaced by a hymn, audible only to him, which sings of a greater harmony, a greater beauty, a greater justice, and of love.
>
> This is the revelation granted to Abram, ... the intuition of something greater than the physical universe he knows, an essence separable from the world and yet eternally involved with it in a continuing state of care and love (Arieti 1981a, pp. 3–4).

The leitmotif of *Abraham and the Contemporary Mind* was the idea of a merging of the physical and the spiritual that entailed not the assimilation of the spirit with the matter but the coexistence of transcendence and immanence—that is, a spiritual dimension within the universe. By acknowledging the existence of "one God, incorporeal, invisible, eternal—transcending any matter" (Arieti 1981a, p. 54), it had been Abraham, well before Plato or Descartes, who gave rise to the dualist tradition that Arieti discussed in the pages of his book:

> The term *dualism* is not agreeable to many people today; they prefer to consider everything that exists in a *monistic* frame of reference. According to the monists, there is no basic difference between mind and organism, psyche and soma, soul and body, the psychological and the physical, being and appearance, God and nature, and so on. ... Whatever the reason, there is no doubt that a monistic vision of the world ... is found by many people to be more appealing. ... Perhaps monism appeals to our desire to interpret the entire universe in one way, supposedly the right one, and to dispense with any apparent mysticism. I myself find that, for reasons I have not totally elucidated but probably have to do with my scientific biological training, I would be inclined to prefer monism. But contrary to the majority of those inclined to monism, I feel we are not in a position to be so. In my view, all attempts to overcome dualism have failed, and dualism is inescapable. We have confused our state of understanding with our wishes (Arieti 1981a, pp. 19, 21–22).

Abraham's intuition was therefore the origin of that "neuropsychiatric split" which still represented an unsolved scientific problem (Arieti 1968). The most refined theoretical attempts to solve it had not succeeded in overcoming "the dualistic barrier" (Arieti 1981a, p. 40). With the concept of isomorphism, for example, von Bertalanffy had suggested the existence of common laws and principles as being at the basis of psychological and physical phenomena. Though this was a tremendous accomplishment, the enigma remained unsolved, for it left the question of the origin of these principles and laws. In a letter sent to von Bertalanffy, upon reading his book on general systems, Arieti addressed this issue:

> Dear Ludwig ..., there is a point in your book which I have not completely understood and about which I would like to ask your explanation. I believe, as you do, that all entities become organized in systems, but I do not understand where you think the organizations or the structures come from. Perhaps you believe that they are immanent in the things themselves, but this is no explanation, because we would still have to explain the immanence of the systematization. The second possibility is that you think the structures and the organizations are transcendent or come from a different universe, but this would be a reaffirmation of Platonism, and I don't believe this is what you imply. The third possibility, of course, requires a theological or mystical intervention and this I don't think is what you mean either. If you have any free time I would appreciate a reply (SAP 1968a).

"You put your finger on some very profound questions where a facile answer would be inappropriate" (SAP 1968b), replied von Bertalanffy. So inappropriate was the answer, in fact, that it couldn't come; the order of things does exist, but why it exists and where it comes from remains unknown.

Is science therefore defeated? With his book on Abraham does Arieti give up his scientific commitment? Does he renounce science for religion, matter for spirit? It does not seem so. In Arieti's view, science inherited a challenge from Abraham—to take charge of the mystery by acknowledging the dualism and the spiritual dimension of life.

John Eccles, whom Arieti had invited to write a chapter for *New Dimensions in Psychiatry* (SAP 1975),[1] also had already headed down this path:

> According to the scientists, the only avenue open to contemporary neurophysiology and psychology is to accept interactionism – that is, the theory of an interaction between psyche and brain. The Nobel Prize winner John E. Eccles, one of the leading neuro-physiologists of our time, believes that dualism and interactionism are evident. In the human being who thinks, feels, assesses ethically, chooses, wills, and loves, neural events do occur, but the human thoughts, feelings, evaluations, choices, and actions are different from the neural events by which they are produced. No "identity theory" is tenable which equates physical events like molecular, biochemical, or electrical phenomena with the psychological events (Arieti 1981a, p. 41).

Thus, science, or at least some scientists, had also started to acknowledge that spiritual dimension of life to which Abraham had drawn the attention.

In 1980, in a lecture at the American Academy of Psychoanalysis entitled "Psychotherapy in a Cultural Climate of Pessimism," Arieti addressed this issue again,

[1] Eccles declined Arieti's invitation, for he was already working on many projects, among which was *The Self and Its Brain,* co-authored with Karl Popper.

stating that contemporary science should not only accept dualism but admit that it actually *needed* it. In a world where the study of the human being tended to become a detached description of biological mechanisms and genetic sequences and where the idea of a predetermined nature tended to prevail, the best defense against the current pessimism hinged on taking charge of the spiritual part of human existence. Therefore, dualism had a hand to play for human science and against scientism (SAP 1980b; Arieti 1981b).

Along with his book on Abraham, "Psychotherapy in a Cultural Climate of Pessimism" can be considered both the spiritual and the scientific testament of Silvano Arieti. In 1980 he started to suffer from ailments that required him to undergo a medical examination (SAP 1980d). It is perhaps for this reason that he did not go to Italy that summer, as is indicated in his correspondence with Luciana, the housekeeper of his summer home in Villasimius (SAP 1980a). In December 1980 he received a letter from Christopher Roberts, a scholar of the musical traditions in primitive cultures, and the letter attached a cartoon of a man sitting at a desk overloaded with books; the man looks thoughtful, his left hand holding up his head and his wide-open eyes looking at a horde of primitive hairy little men with bones through their noses. Behind the men are some thatched sheds, and at the center of the horde a taller man playing a contrabass. The caption under the drawing reads "Dear Dr. Arieti, get well soon and write a book about all this" (SAP 1980f).

In January 1981, Arieti returned to Italy for the last time. He traveled to Pisa, where he was given a splendid welcome. "I thought how proud your parents would have been of you, if only they could see that crowded room" (SAP 1981a, my translation), his cousin Lydia wrote. Then, in April, just a few months before his death, he visited Israel, where he delivered a speech at the Museum of Jewish Diaspora of Tel Aviv. He was accompanied by his wife Marianne, his Aunt Yolanda, his uncle Giampaolo, and his cousin Lydia (SAPb; SAP 1980e, 1981b).[2]

He courageously faced his terminal cancer with full awareness (Clemmens, Spiegel, Bieber, & Di Cori 1982). He did not cease to look outside, observing what surrounded him, worrying about the world he was about to leave. As a Democrat, he was concerned for the America under President Ronald Reagan (SAP 1980c). At the same time, he did not abandon his attitude of looking inwardly, perhaps more a part of his nature than a consequence of his profession. In April 1981 he admitted to his cousin Rabbi Jack Bemporad that he had begun undergoing psychoanalysis again, for he felt that he still had to work on himself, and he hoped it was not too late (SAP 1981c). Thus, for his whole life Arieti put into practice the idea that "until his last breath, the human being can grow, can innovate, and to some extent, with the help of others, he can recreate his life" (Bemporad 1981, p. vii).

His search for self-awareness, his firm belief that one can always choose the way to deal with the vicissitudes of life, his endeavor to face even the most difficult constraints as a free human being ended with his "serious smile" (Clemmens, Spiegel, Bieber, & Di Cori 1982, p. 9) on August 7, 1981, when at age sixty-seven he

[2] Arieti had been invited by Mauro Curradi, Director of the Institute of Italian Culture of Tel Aviv.

surrendered to cancer. "We have witnessed a stellar event," his analyst Rose Spiegel remarked after his death (Clemmens, Spiegel, Bieber, & Di Cori 1982, p. 8).

In 1967 Silvano Arieti, referring to the work of Giambattista Vico, had described a human being as "a finite center of possibility that tends to the infinite" (Arieti 1967, p. 148). These words seem to describe also Arieti's life and his conception of science as an ongoing search for knowledge "to break the secret of the universal night and make a piece of understanding a piece of ourselves" (Arieti 1967, p. 453).

References

Arieti, S. (1961). A Re-Examination of the Phobic Symptoms and of Symbolism in Psychopathology. *The American Journal of Psychiatry,* 118 (2), 106–110.
Arieti, S. (1967). *The Intrapsychic Self.* Basic Books.
Arieti, S. (1968). The Present Status of Psychiatric Theory. *The American Journal of Psychiatry,* 124 (12), 1630–1639.
Arieti, S. (1971). Psychodynamic Search of Common Values with the Schizophrenic. In *International Congress Series N. 259. Psychotherapy of Schizophrenia. Proceedings of the IV International Symposium, Turku, Finland, August 4–7* (pp. 94–100). Excerpta Medica.
Arieti, S. (1972). *The Will To Be Human.* Quadrangle Books.
Arieti, S. (1979a). New Views on the Psychodynamics of Phobias. *The American Journal of Psychotherapy,* 33 (1), 82–95.
Arieti, S. (1979b). *The Parnas.* Basic Books.
Arieti, S. (1980). Man's Spirituality and Potential for Creativity as Revealed in Mental Illness. *Comprehensive Psychiatry,* 21 (6), 436–443.
Arieti, S. (1981a). *Abraham and the Contemporary Mind.* Basic Books.
Arieti, S. (1981b). Psychoanalytic Therapy in a Cultural Climate of Pessimism. *The Journal of the American Academy of Psychoanalysis,* 9 (1), 171–184.
Bemporad, J. (1981). In Memoriam. Silvano Arieti 1914–1981. *Journal of the American Academy of Psychoanalysis* 9 (4), III–VII.
Clemmens, E., Spiegel, R., Bieber, I., & Di Cori, F. (1982). Silvano Arieti: 1914–1981. *Academy Forum,* 26 (1) 6–9.
Forti, C. (1998). *Il caso Pardo Roques. Un eccidio del 1944 tra memoria e oblio.* Einaudi.
Gardner, S. (1979). Fascist-Holocaust Survivor, Author, to Open Sephardic Program Sept.9. *The Jewish Week,* 192 (11), 22.
Gerchick, R. (1979, October 18). The Parnas. *Scarsdale Inquirer,* 8.
Moravia, S. (1986). Per una razionalità plurale. La nuova episteme e la scienza dell'uomo. In G. Invitto (Ed.), *Il pungolo dell'umano. Conversazione su un impegno filosofico* (pp. 92–117). Franco Angeli.
SAPa. *Speeches and Writings, 1940–1981. The Parnas. Correspondence, 1976–198o.*
SAPb. *Speeches and Writings, 1940–1981. The Parnas. Italian Cultural Institute, 1980–1981.*
SAPc. *Speeches and Writings, 1940–1981. The Parnas. Promotion, 1979.*
SAPd. *Speeches and Writings, 1940–1981. The Parnas. Reviews, 1979–1981.*
SAPe. *Subject File, 1914–1981. Education. Workbooks, circa 1930.*
SAP (1965). L, June 3 (US to Silvano Arieti). In *Correspondence 1940–1981.*
SAP (1968a). L, March 18 (Silvano Arieti to Ludwig von Bertalanffy).
SAP (1968b). L, October 30 (Ludwig von Bertalanffy to Silvano Arieti).
SAP (1971). T (Silvano Arieti, *Revisiting the Concept of Transference and Counter-Transference in the Psychoanalytic Therapy of Schizophrenia*). In *Speeches and Writings, 1940–1981. Lectures.*
SAP (1975). L, February 25 (John Eccles to Silvano Arieti). In *Correspondence, 1940–1981.*

SAP (1976). L, September 7 (Silvano Arieti to Francis Braceland). In *Correspondence, 1940–1981.*
SAP (1978). T, September 22 (Ferruccio Di Cori, *Introductory Speech for the Presentation of the Sigmund Freud Award to Silvano Arieti By the American Association of Psychoanalytic Physicians*). In *Speeches and Writings, 1940–1981. Lectures.*
SAP (1979a). L, January 25 (Primo Levi to Silvano Arieti). In *Correspondence, 1940–1981.*
SAP (1979b). L, February 21 (Primo Levi to Silvano Arieti). In *Correspondence, 1940–1981.*
SAP (1979c). L, May 28 (Silvano Arieti to Peter Yarrow). In *Speeches and Writings, 1940–1981. The Parnas. Promotion, 1979c.*
SAP (1979d). L, October 31 (Mary Higgins to Silvano Arieti). In *Speeches and Writings, 1940–1981. The Parnas. Promotion, 1979d.*
SAP (1979e). L, November 22 (Primo Levi to Silvano Arieti). In *Correspondence, 1940–1981.*
SAP (1980a). L, April 4 (Ms. Luciana to Silvano Arieti). In *Correspondence, 1940–1981.*
SAP (1980b). T, May 4 (Silvano Arieti, *Psychoanalytic Therapy in a Cultural Climate of Pessimism*). In *Speeches and Writings, 1940–1981. Lectures.*
SAP (1980c). L, May 21 (Silvano Arieti to Senator Henry Jackson). In *Correspondence, 1940–1981.*
SAP. (1980d). L, June 3 (Silvano Arieti to Perry Fersko). In *Correspondence, 1940–1981.*
SAP (1980e). L, December 11 (Mauro Curradi to Silvano Arieti). In *Speeches and Writings, 1940–1981. The Parnas. Italian Cultural Institute, 1980e–1981.*
SAP (1980f). L, December 18 (Christopher Roberts to Silvano Arieti). In *Correspondence, 1940–1981.*
SAP (1981a). L, February 21 (Lydia Bemporad to Silvano Arieti). In *Correspondence, 1940–1981a.*
SAP (1981b). L, March 2 (Silvano Arieti to Mauro Curradi). In *Speeches and Writings, 1940–1981b. The Parnas. Italian Cultural Institute, 1980–1981b.*
SAP (1981c). L, April 6 (Silvano Arieti to Jack Bemporad). In *Correspondence, 1940–1981c.*
Seebohm, C. (1979, September 23). The Parnas. *The New York Times*, 14

Index

R. Passione, *Psychiatry and the Human Condition*, Springer Biographies,
https://doi.org/10.1007/978-3-031-09304-3

C

N

O

P